T0344890

Photovoltaic Sources Modeling

Photovoltaic Sources Modeling

Giovanni Petrone
University of Salerno, Italy

Carlos Andrés Ramos-Paja
National University of Colombia

Giovanni Spagnuolo
University of Salerno, Italy

Registered Offices
John Wiley & Sons, Inc., 111 River Street, Hoboken, NJ 07030, USA
John Wiley & Sons Ltd, The Atrium, Southern Gate, Chichester, West Sussex, PO19 8SQ, UK

Editorial Office
The Atrium, Southern Gate, Chichester, West Sussex, PO19 8SQ, UK

For details of our global editorial offices, customer services, and more information about Wiley products visit us at www.wiley.com.

Wiley also publishes its books in a variety of electronic formats and by print-on-demand. Some content that appears in standard print versions of this book may not be available in other formats.

Library of Congress Cataloging-in-Publication Data

Names: Petrone, Giovanni, author. | Ramos-Paja, Carlos Andrés, author. | Spagnuolo, Giovanni, author.
Title: Photovoltaic sources modeling / Giovanni Petrone, University of Salerno, Italy, Carlos Andrés Ramos-Paja, Universidad Nacional de Colombia, Giovanni Spagnuolo, University of Salerno, Italy.
Description: Chichester, West Sussex, United Kingdom : John Wiley & Sons, Inc., [2017] | Includes bibliographical references and index.
Identifiers: LCCN 2016038240 (print) | LCCN 2016054257 (ebook) | ISBN 9781118679036 (cloth) | ISBN 9781118756126 (pdf) | ISBN 9781118756492 (epub)
Subjects: LCSH: Photovoltaic power generation–Mathematical models.
Classification: LCC TK1087 .P43 2017 (print) | LCC TK1087 (ebook) | DDC 621.31/244011–dc23
LC record available at https://lccn.loc.gov/2016038240

Cover image: Tetra Images/Gettyimages
Cover design by Wiley

Set in 10/12pt Warnock by SPi Global, Pondicherry, India
Printed and bound in Malaysia by Vivar Printing Sdn Bhd

10 9 8 7 6 5 4 3 2 1

To Patrizia, Nicola and Marco Andrea
Giovanni Petrone

To Claudia Patricia and Alejandro
Carlos Andrés Ramos-Paja

To Simonetta and Valeria
Giovanni Spagnuolo

Contents

Acknowledgements

The authors wish to acknowledge the research groups they belong to, the University of Salerno and the Universidad Nacional de Colombia, for the moral and technical support in achieving the research results and the book writing.

G. Petrone and G. Spagnuolo also acknowledge the University of Salerno and the Italian Ministry of Education for having funded, through many projects, the research activities the results of which are outlined in this book.

C.A. Ramos-Paja also acknowledges the Universidad Nacional de Colombia and Colciencias (Fondo Nacional de Financiamiento para Ciencia, la Tecnología y la Innovación Francisco José de Caldas) for having funded project MicroRENIZ-25439 (Code 1118-669-46197) the results of which are outlined in this book.

G.Petrone and G.Spagnuolo acknowledge Prof. N.Femia and Prof.L.Egiziano, from the University of Salerno, for the joint research activity in photovoltaics over the last sixteen years and for their support since the PhD course studies.

SolarTechLab[1] of Politecnico di Milano (Italy) is acknowledged for some photos in Chapter 4.

The contribution of Dr. Martha Lucia Orozco Gutierrez of the Universidad del Valle (Colombia) is warmly acknowledged for her contributions to Chapters 5 and 6.

1 http://www.solartech.polimi.it.

Introduction

The idea of writing this book was born ten years ago, when the authors, who before then used the simple electrical models that had been reported in the – very scant – literature at that time, published their first paper on this topic. At that time, the values of the parameters of the widely used single-diode circuit model (SDM) were identified through approximate procedures, often consisting of two steps, the first neglecting the loss mechanisms and the following one introducing them by fixing the values of the other parameters. Mismatching phenomena, and especially partial shadowing, were accounted for by simple multiplication of the array power production in standard test conditions by a loss factor. Events producing degradation of the cells and bypass diodes were only observed or simulated in simplified cases. Also, the need to simulate the array, including bypass and blocking diodes, in unconventional operating conditions and in real time using an embedded system, did not arise at that time, so that the problem of having a fast, accurate, efficient and reliable simulation of even large arrays was treated in literature only marginally. Ten years later, the authors have gained an improved knowledge of these matters, which has been condensed in this book with the main aim of driving the reader towards the best and more suitable array model for a given application.

The SDM is used first, in Chapter 1, for fast simulations of even large arrays working in standard conditions through a suitable symbolic manipulation of the non-linear equation describing it at the terminals through the current and voltage variables. This chapter gives the background for understanding almost all the advanced approaches presented in the book. In order to use the aforementioned model in practice, and also the others in the book, the reader has to turn the data available in the datasheet provided by any panel manufacturer into the parameters involved into the circuit model. This problem is addressed in the second chapter of the book, where methods – both approximations using explicit formulae and accurate ones – requiring the solution of a non-linear system of equations, are compared. The SDM presented in the first chapter and the parameters values determined by the methods described in Chapter 2 are thus ready for simulating arrays working in the so-called "standard test conditions" the datasheet measurements refer to. Chapter 3 allows the reader to extend the simulation capabilities of the SDM by suggesting a number of additional equations allowing simulation of the array in any operating conditions, both in terms of irradiance and temperature values, under the assumption that all the cells in the array work in exactly the same operating conditions and are described by the SDM with the same parameter values.

Mismatched conditions, and especially the presence of the partial shadowing phenomenon affecting the array, need more sophisticated simulation models and tools. Chapter 4 gives an overview of the main sources of degradation and mismatching, pointing out that in many cases the effect is at cell level, and so the modeling of such conditions may require models with different levels of detail and granularity. Chapter 5 introduces a first possibility, describing a computationally efficient simulation model that works at the granularity level of the module. Thus each panel can be simulated by accounting for up to two or three different operating conditions affecting the modules from which it is made. The proposed approach allows for very fast simulation of the mismatched array, with more accuracy than when introducing a simple power reduction. Nevertheless, cell-level phenomena such as hot spots and a detailed string behavior can be achieved only by modeling the array at cell level. For fast simulation of arrays consisting of thousands of cells, this requires more detailed models and the use of algorithmic approaches, and also the consideration of cell reverse biasing.

Chapter 6 gives the modeling and algorithmic recipes for approaching this task in a proper way: some examples and comparisons are presented in order to show the reader that a suitable organization of the equations in the overall non-linear model allows for a saving of hours of simulation time compared to the use of the standard algorithms that are built into commercial mathematical software packages. Once the reader has chosen the right array model for the analysis that he or she has to perform, it is merged with the model of the power processing system that allows harvesting of the maximum PV power at any given operating condition. Indeed, the literature and market trends clearly indicate that, in the near future, electronics will be embedded into PV modules with the aim of maximizing the power output and to make its operating point independent of other modules connected in series or in parallel to it. Thus the module and its dedicated electronics will no longer be distinguishable and their integrated modeling will be required.

Chapter 7 shows that a suitable array model is needed for the design of the switching converter that controls the power flux. Static and dynamic performance can be optimized provided that a suitable simulation of the whole system, including both the array and the power-processing system, is performed. Once this has been achieved, different control techniques can be used, as summarized in Chapter 8.

Tables of Symbols and Acronyms

Table 1 Acronyms

Acronym	Definition
DDM	Double-diode model
DMPPT	Distributed maximum power point tracking
FF	Fill factor
GCR	Ground-cover ratio
ISDM	Ideal single-diode model
I–V	Current vs voltage
MIU	Module integrated unit
MPP	Maximum power point
MPPT	Maximum power point tracking
P–V	Power vs voltage
PV	Photovoltaic
P&O	Perturb and observe MPPT algorithm
SDM	Single-diode model
SIF	Shade-impact factor
SSDM	Simplified single-diode model
STC	Standard test conditions

Table 2 Symbols

Symbol	Definition	Value or units
C_0	Temperature coefficient	A/K^{-3}
E_g	Silicon energy gap	J
G	Irradiance	W/m^2
G_0	Irradiance reference condition	W/m^2
G_{STC}	Irradiance in STC	$1000\ W/m^2$
I_{MPP}	Current at the maximum power point	A
i_{MPP}	Small signal current at the maximum power point	A
I_{ph}	PV photo-induced current	A
I_{sc}	PV short-circuit current	A
I_s	PV diode saturation current	A
$I_{s,db}$	Bypass diode saturation current	A
$I_{s,blk}$	Blocking diode saturation current	A
I_{STC}	DC current in STC	A
K	Boltzman constant	1.38×10^{-23} J/K
$M(D)$	Voltage conversion ratio of a power converter	n/a
P_{MPP}	Power at the maximum power point	W
q	Electron charge	1.6×10^{-19} C
R_s	Series resistance	Ω
R_h	Shunt resistance	Ω
T	Cell temperature	K
T_0	Cell temperature reference condition	K
T_a	Ambient temperature	K
T_m	Temperature at the backside of the PV modules	K
T_{STC}	Cell temperature in STC	298K
V_{MPP}	Voltage at the maximum power point	V
v_{MPP}	Small signal voltage across the maximum power point	V
V_{oc}	PV open-circuit voltage	V
V_{STC}	DC voltage in STC	V
V_t	Thermal voltage of PN junction	V
$V_{t,db}$	Thermal voltage of bypass diode	V
$V_{t,blk}$	Thermal voltage of blocking diode	V
$W(\theta)$	Lambert W function with argument θ	n/a
α_i	Temperature coefficient of the short circuit PV current	(%)/°C
α_v	Temperature coefficient of the open circuit PV voltage	(%)/°C
η	Diode ideality factor	n/a
η_{pc}	Efficiency of a power converter	n/a

1
PV Models

1.1 Introduction

As for any physical system, PV cell modeling can be done with different levels of accuracy, depending on the user's purposes. In this book the PV generator is always modeled through an equivalent circuit and by using concentrated parameters and variables. Indeed, the aim is to give the reader tools for implementing the PV array model in simulation environments, allowing them to analyze and design the whole PV generator, including the power-processing system feeding a load or the grid. For users interested in this kind of study, access to data concerning the physical properties of the semiconductor material involved or parameter values that depend on the cells' manufacture, such as dopant concentrations or material response to the radiation spectrum, is either not easy or the figures are not readily translated into circuit parameters. Instead, the use of laboratory measurements at the PV terminals, in terms of current and voltage values, or use of experimental data from the product datasheet, are more viable ways of proceeding.

This chapter introduces the two main circuit PV models used in the literature: the single-diode and the double-diode models. The first is the more widely used because of the reduced number of circuit parameters to be identified. The double-diode model (DDM) has better accuracy, especially at low irradiance levels, but it requires a more involved identification of the parameter values. Thus while some space is dedicated in this chapter to the DDM, in the following chapters the single-diode model (SDM) will be considered the reference one.

1.2 Modeling: Granularity and Accuracy

In the last five years a huge literature has been devoted to modeling PV sources, for two main purposes. The first is the reproduction of the I–V curve at the generator terminals through a suitable electrical model, regardless of the size of the PV source: from one PV cell, or even sub-portions of it, up to large PV fields made of series-connected modules forming strings that are in turn connected in parallel. The second purpose is performing energetic analyses concerning plant productivity, using models based on empirical or semi-empirical equations.

The first set of models aims to describe the functional current–voltage relationship at the PV terminals on the basis of the equivalent electrical circuit of the PV source.

Photovoltaic Sources Modeling, First Edition. Giovanni Petrone, Carlos Andrés Ramos-Paja and Giovanni Spagnuolo.

Companion Website: www.wiley.com/go/petrone/Photovoltaic_Sources_Modeling

Such models are usually scalable and the parameter values can be varied according to the operational weather conditions at the PV source. They are also useful for modeling unusual operating conditions, such as mismatches due to partial-shading phenomena. The implementation in circuit-oriented simulators such as PSPICE and PSIM, or in general-purpose simulation environments such as MATLAB® and SCILAB, is almost straightforward. This allows study of the PV source together with dedicated controls, power-processing systems, switching converters, and so on. The PV-source non-linearity is accurately reproduced by these models, but at the cost of a significant computational burden, especially if the granularity of the simulation is at the cell, or even sub-cell level, in the context of a large PV field consisting of thousands of cells. As a consequence, PV equivalent circuit models allow simulations of systems to be performed over short time windows, usually of fractions of a minute.

The second set of models is aimed at performing long-term analyses – over days or months – and they therefore cannot use detailed descriptions of the current–voltage relationship in any operating conditions. Instead, they use simplified equations in which the energy produced by the PV source is described as a function of the environmental conditions and parameters related to the installation type. These approaches usually require tuning, with correcting factors based on experimental measurements used to account for the real operating conditions of the PV field. Approaches based on fuzzy logic [1] and neural networks [2] are often used to take into account the historical meteorological data of the installation site when estimating the produced energy, in order to predict the pay-back time and to evaluate the economic viability of the PV system.

The largest part of this book is focused on the first set of models. The next sections introduce the main circuit models used for PV source simulation in ordinary operating conditions.

1.3 The Double-diode Model

The double-diode model (DDM) describes the PN junction operation through the Shockley equation, and includes series and shunt resistances to incorporate the current-dependent and the voltage-dependent loss mechanisms. It has the important feature of modeling the carrier-recombination losses in the depletion region.

The non-linear equation giving the relationship between the current and the voltage at the PV source terminals is:

$$I_{pv} = I_{ph} - I_{s1}\left[e^{\frac{(V_{pv}+I_{pv}R_s)}{\eta_1 V_t}} - 1\right] - I_{s2}\left[e^{\frac{(V_{pv}+I_{pv}R_s)}{\eta_2 V_t}} - 1\right] - \frac{V_{pv}+I_{pv}R_s}{R_h} \tag{1.1}$$

This equation can be represented in any circuit-oriented simulator by the equivalent circuit shown in Figure 1.1. I_{d1} and I_{d2} are the second and the third additive terms in (1.1).

The photoinduced current I_{ph}, the two diodes' saturation currents I_{s1} and I_{s2}, the two diodes' ideality factors η_1 and η_2, and the two resistances R_s and R_h are unknown parameters. Their values have to be identified on the basis of measurements performed by the producer or by the user, in the laboratory or in the field, on the real PV unit that is to be simulated. $V_t = \frac{kT}{q}$ is the thermal voltage of the PN junction in the PV cell, and

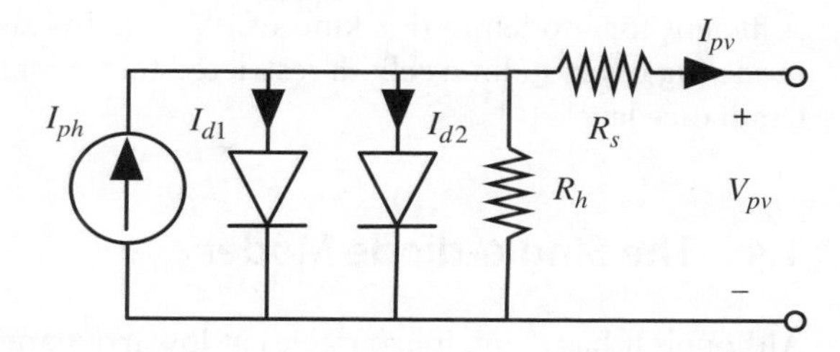

Figure 1.1 Equivalent circuit of the double-diode model.

it is calculated by considering the cell temperature. The value of the second saturation current I_{s2} is three to five times higher than the first:

$$I_{s2} = \frac{T^{\frac{2}{5}}}{3.77} \cdot I_{s1} \tag{1.2}$$

Detailed expressions of the two saturation currents, including the temperature dependencies, are given in the literature [3, 4]. As for the two diodes' ideality factors, the first is often given in the literature at a value of 1 and the second at 2 [4], although their ranges of variation are $\eta_1 \in [1, 1.5]$ and $\eta_2 \in [2, 5]$ [5]. Such findings, related to the physical properties of the semiconducting material, allow a reduction in the number of parameter values to be identified on the basis of measurements performed on the specific PV source that is being modeled. This means that a small set of measurements and data, sometimes even just those given in the manufacturer's datasheet, can be used for the identification process.

Hejri et al. scaled down the number of parameter values to be identified from seven to five by fixing $\eta_1 = 1$ and $\eta_2 = 2$ [6]. This allowed them to use data available in the PV module or cell datasheet, so the problem became similar to using the SDM (see Section 2.2.1).

Besides the aforementioned one, other approaches in the literature are always based on a reduction of the number of parameter values to be identified because of the lack of seven significant sets of experimental data. For instance, the loss mechanisms might be neglected and the relationship between the two saturation currents might be used for constructing an iterative procedure converging to the optimal fitting values of the remaining parameters [7].

Such approaches, based on a preliminary symbolic manipulation of the equations, on fixing the value of some of the parameters at values taken from the literature, or on neglecting some of the terms, contrast with the adoption of an accurate model like the DDM. In the rigorous case, when all the values of the seven parameters occurring in the DDM have to be identified, fitting methods based on optimization algorithms, often based on stochastic approaches, need to be used [8]. They require large computational resources and contact with the physical problem under examination is lacking.

The DDM is able to describe accurately the behavior of thin-film cells because of its ability to give a smoother curve in the region of the maximum power point (MPP). Indeed, a high value of the ratio $\frac{I_{s2}}{I_{s1}}$ matches the typical I–V curve of thin-film cells, while $I_{s2} = 0$ gives a more abrupt transition from the constant-current to the constant-voltage branches of the I–V curve, which is much more typical of crystalline-cell technology [9]. This is confirmed by literature data and studies on crystalline-Si cells, which show that junction recombination is a mechanism that can be neglected, so that a SDM is sufficient for modeling this kind of PV cell. The usefulness of the double-diode-based modeling of crystalline cells is restricted to an improved accuracy at both high and low irradiance levels [4].

1.4 The Single-diode Model

Although it has some inaccuracies at low irradiance levels, the SDM is very often used in the literature to model crystalline-silicon-based PV generators. The recombination current is neglected; it is fixed at $I_{s2} = 0$ in (1.1) so that the corresponding equation is given by (1.3). The simplified equivalent circuit is shown in Figure 1.2.

$$I_{pv} = I_{ph} - I_s \left[e^{\frac{(V_{pv}+I_{pv}R_s)}{\eta V_t}} - 1 \right] - \frac{V_{pv} + I_{pv}R_s}{R_h} \tag{1.3}$$

Depending on the desired simulation accuracy, the SDM can be simplified further. In the literature it is often considered a lossless model; see Figure 1.3a, where both the series and the shunt resistances are neglected, the former put to zero and the latter to infinity. This choice leads to inaccuracies:

- when the numerical analysis involves very high irradiance levels, because the significant voltage drop across R_s is neglected (see Figure 1.3b)
- in mismatched conditions, because the significant current flowing into R_h in these conditions is not modeled (see Figure 1.3c).

The non-linear functions in (1.4)–(1.6) give the current–voltage relationships corresponding to the equivalent circuits shown in Figures 1.3a–c, respectively.

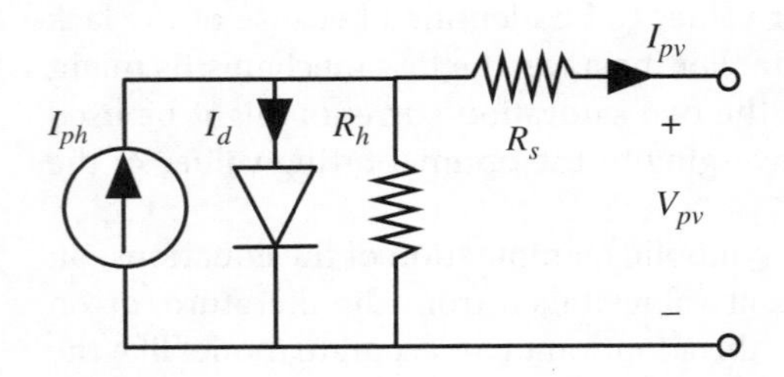

Figure 1.2 Equivalent circuit of the single-diode model.

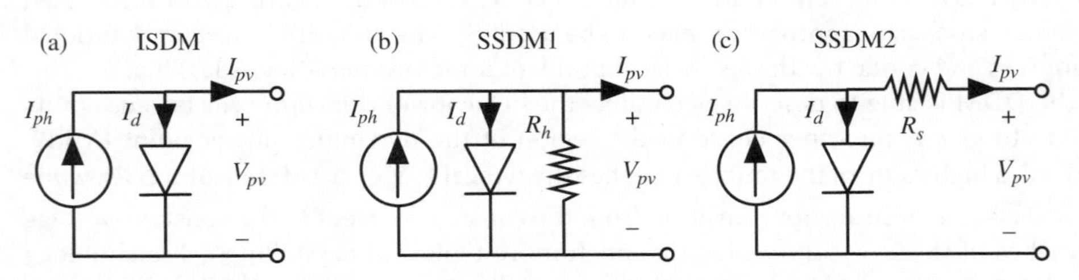

Figure 1.3 Equivalent circuits of (a) the ideal single-diode model; (b) and (c) two simplified single-diode models.

$$I_{pv} = I_{ph} - I_s \left(e^{\frac{V_{pv}}{\eta V_t}} - 1 \right) \tag{1.4}$$

$$I_{pv} = I_{ph} - I_s \left(e^{\frac{V_{pv}}{\eta V_t}} - 1 \right) - \frac{V_{pv}}{R_h} \tag{1.5}$$

$$I_{pv} = I_{ph} - I_s \left[e^{\frac{V_{pv}+I_{pv}R_s}{\eta V_t}} - 1 \right] \tag{1.6}$$

1.4.1 Effect of the SDM Parameters on the I–V Curve

Equation 1.3, or equivalently its approximate versions, allows us to know the I–V relationship for any PV source operating in homogeneous conditions. Indeed, under the assumption that all the cells are equal and subject to the same environmental conditions, the parameters (I_{ph}, I_s, η, V_t, R_s, R_h) can be scaled up or down by accounting for the number of the cells/panels connected in series to form a string, and the number of strings connected in parallel. In such conditions the voltages are multiplied by N_s, the number of the series-connected cells, and all the currents are multiplied by N_p, the number of parallel-connected strings. Consequently, the series resistance R_s and the parallel resistance R_h increase by a factor N_s and are divided by a factor N_p. The SDM parameters are scaled as follows:

$$I_{ph} = N_p \cdot I_{ph,cell} \tag{1.7}$$

$$I_s = N_p \cdot I_{s,cell} \tag{1.8}$$

$$\eta = \eta_{cell} \tag{1.9}$$

$$V_t = N_s \cdot V_{t,cell} \tag{1.10}$$

$$R_s = \frac{N_s}{N_p} \cdot R_{s,cell} \tag{1.11}$$

$$R_h = \frac{N_s}{N_p} \cdot R_{h,cell} \tag{1.12}$$

An example of the use of these scaling equations is reported in Table 1.1, where the SDM parameters of a crystalline PV panel operating in standard test conditions (STC) have been derived starting from the corresponding cell parameters. Figure 1.4 shows the corresponding I–V curve (continuous line), obtained using the scaled parameters in (1.3).

Table 1.1 Scaling up the SDM parameter for a PV panel of 36 cells in series.

Parameters	Units	Cell values	Panel values
I_{ph}	A	6.1	6.1
I_s	A	1.2172×10^{-9}	1.2172×10^{-9}
R_s	mΩ	9.2	331.2
R_p	Ω	12.7	457.2
η		1.1377	1.1377
V_t	mV	25.7	924.1

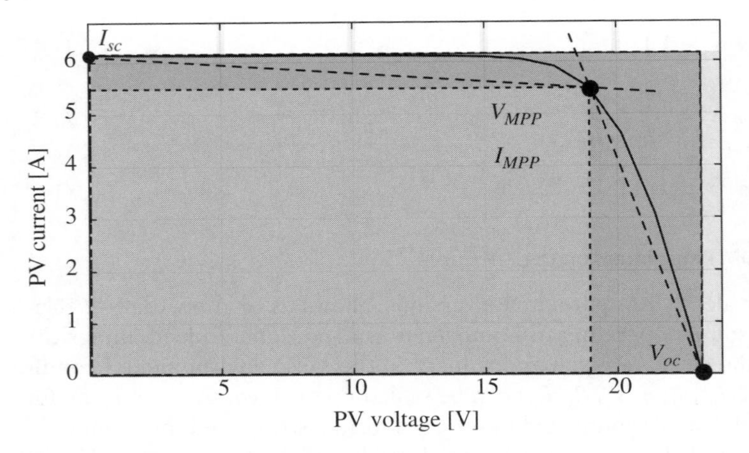

Figure 1.4 Fill factor identification on a I–V curve.

For each PV source, the three main operating points are the ones marked on the I–V curve of Figure 1.4: the short-circuit current (I_{sc}), the MPP (V_{MPP}, I_{MPP}), and the open-circuit voltage (V_{oc}). The fill factor (FF) is the basic indicator usually used for quantifying the quality of a PV panel. The FF is defined as in (1.13); it represents the ratio between the areas of the two rectangles shown in Figure 1.4. The higher the FF, the lower is the slope of the I–V curve in the region of the short-circuit current and the higher is the slope close to the open-circuit voltage.

$$FF = \frac{V_{MPP} \cdot I_{MPP}}{V_{oc} \cdot I_{sc}} \tag{1.13}$$

Figures 1.5 and 1.6 show the modification of the I–V curve due to SDM parameter variations. The photoinduced current (I_{ph}) affects mainly the vertical translation of the I–V curve, while the saturation current (I_s) and the ideality factor (η) affect the horizontal translation of the I–V curve. As will be shown in Chapter 3, such I–V curve translations are directly related to the irradiance (G) and ambient-temperature (T_a) operating conditions of the PV panel under study. The values of R_s and R_h affect mainly the slopes of the I–V curve in the regions of the open-circuit voltage and short-circuit current respectively. The variation of these two parameters is mainly related to degradation and aging, which affects the performance of the PV panel.

1.5 Models of PV Array for Circuit Simulator

From the electrical point of view, the PV source might be reproduced in any circuit simulator using linear components and a diode that represents the PN junction inside the PV cells, as shown in Figure 1.7. This representation is suitable for simulating the PV system in stationary environmental conditions because all parameters have been assumed constant. Nevertheless, in many cases it is desirable to simulate the PV source in the presence of irradiance and temperature variations, for example for testing maximum power point tracking (MPPT) algorithms. This necessitates many changes to the parameters $\overline{\mathbf{P}} = [I_{ph}, I_s, \eta, R_s, R_h]$ during the simulations on the basis of the environmental variations.

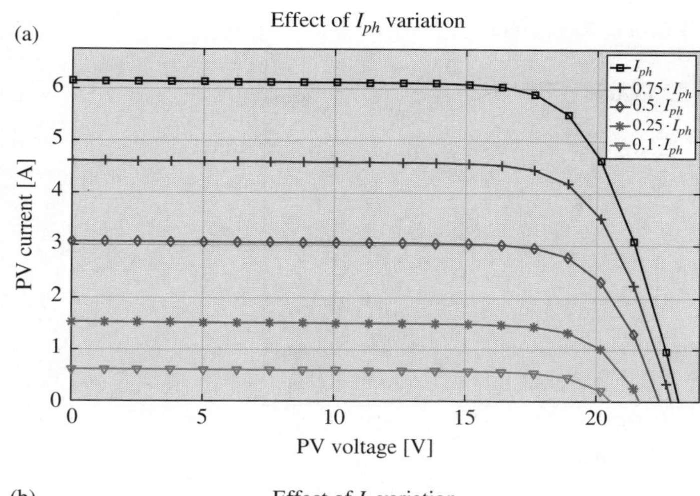

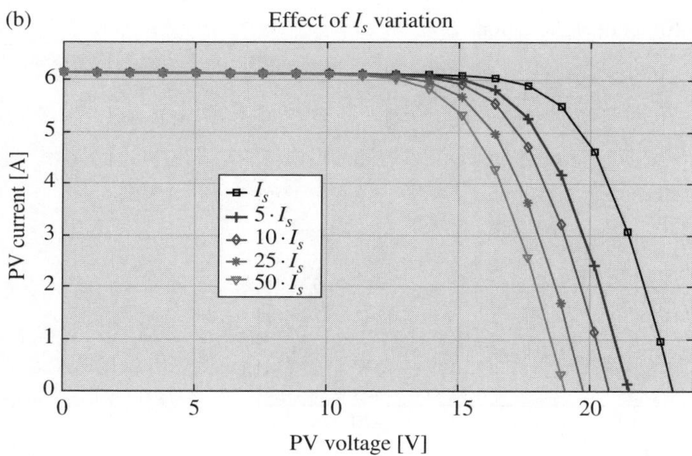

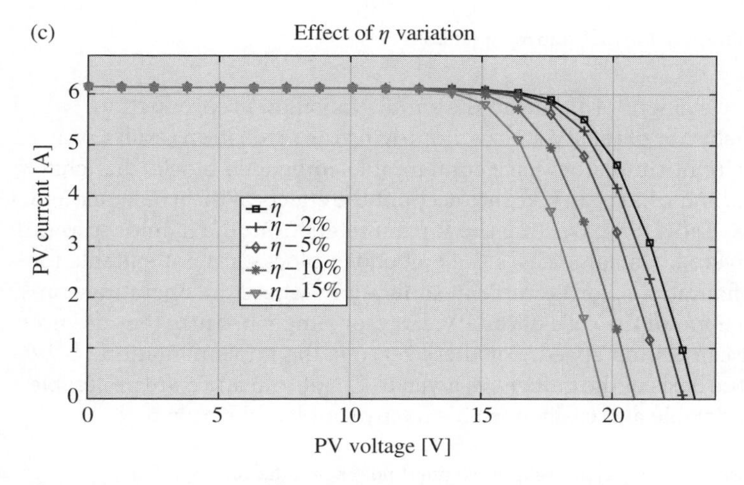

Figure 1.5 I–V curve modification due to SDM parameter variation.

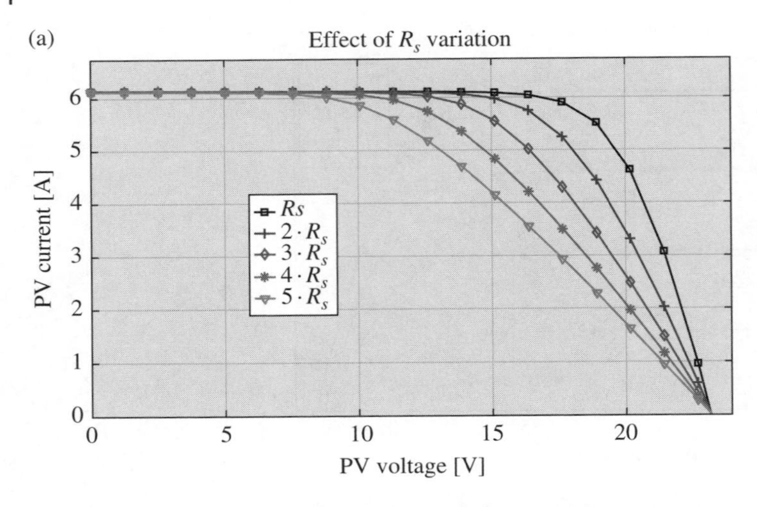

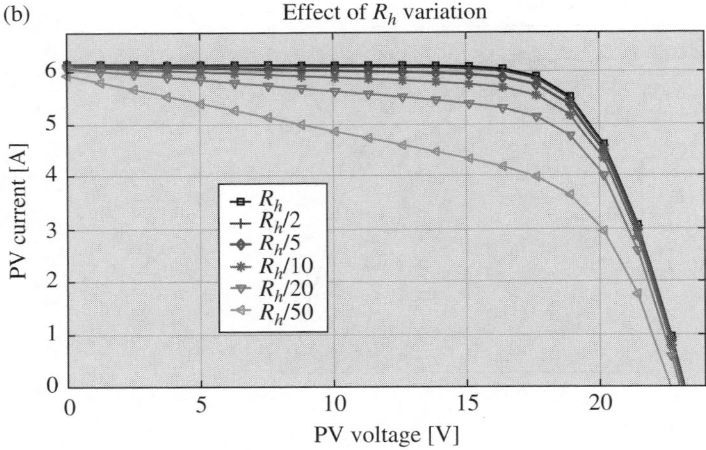

Figure 1.6 I–V curve modification due to R_s and R_h variation.

The modification of any element of $\overline{\mathbf{P}}$ during the simulation allows reproduction, with high precision, of the behavior of the PV source, improving the simulation results. Many commercial software circuit simulators have configurable embedded blocks that characterize the PV sources, and which can take into account the effects of the irradiance and temperature variations. Tools for extracting the $\overline{\mathbf{P}}$ parameters from the manufacturers' datasheet are also delivered.[1] In some cases the embedded blocks are not suitable for studying complex configurations, for example in simulations where the operating conditions are not the same for all the cells of the PV array, or simply because they are not optimized in terms of calculation speed. For these reasons the implementation of PV models developed on the basis of the guidelines given in Chapters 5 and 6 are preferable because they are more flexible and easily portable to any simulation software.

1 The reader is invited to visit, for example, the webpage at: http://powersimtech.com/solutions/renewable-energy/.

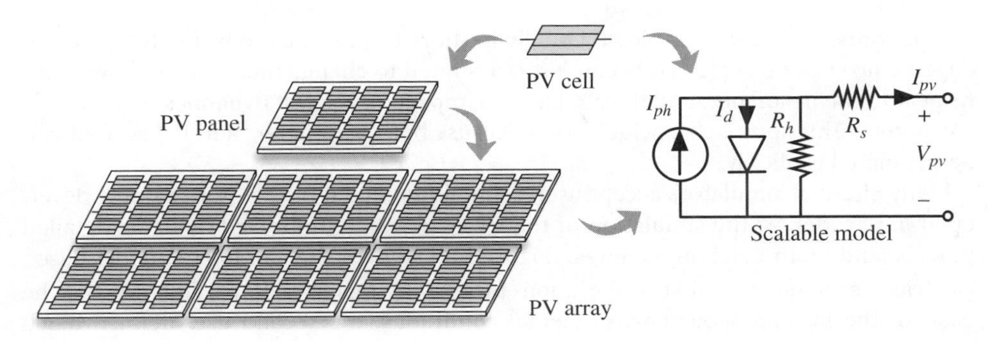

Figure 1.7 The scalable SDM can be used to model a single PV cell or a large PV field operating in homogeneous and stationary environmental conditions.

If the simulator allows implementation of mathematical equations, the PV sources can be reproduced using the scheme shown in Figure 1.8a, where the parameters η, I_{ph}, and I_s are calculated as functions of the instantaneous irradiance and temperature conditions. The parameters R_s and R_h are assumed to be constant. The voltage (V_d) across the parallel resistance R_h is measured at each simulation step, so the current (I_g) is calculated explicitly by exploiting the SDM non-linear equation, as shown in the figure. The controlled current generator is used to interface the numerical part with the electrical

Figure 1.8 PV electrical schemes based on computational blocks suitable for simulating PV fields in presence of irradiance (G) and ambient temperature (T_a) variations.

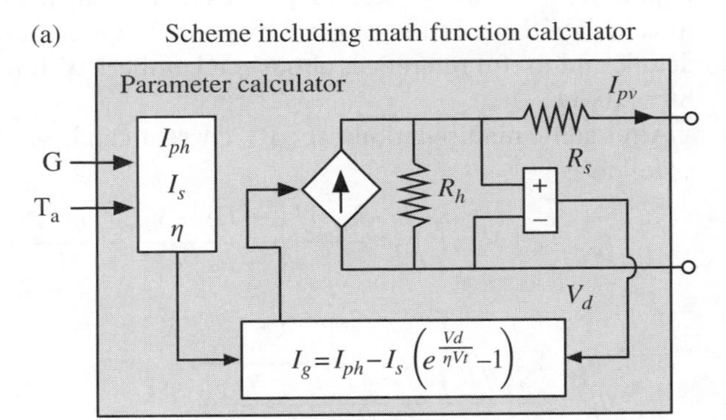

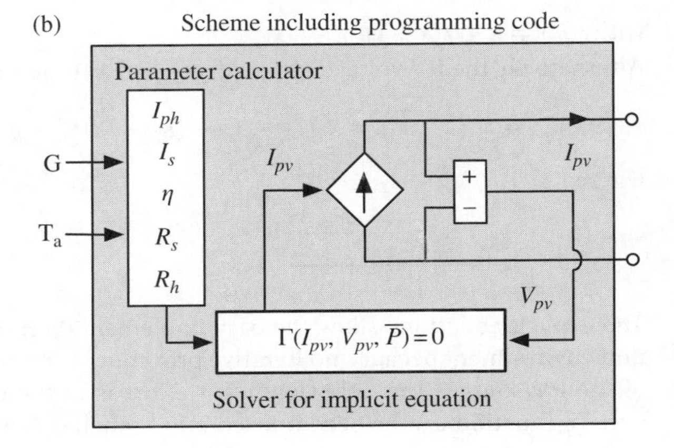

circuit connected to the PV block. Usually in short-term simulations, the temperature dependence can be neglected because it is assumed to change much more slowly with respect to the irradiance, and thus to have no impact on the fast dynamics of the whole PV system. This approach is widely used because both parameters and PV current can be calculated explicitly.

Many circuital simulators accept user-defined functional blocks (for example developed in C code), so the simulation of the PV field can be improved by using detailed models and optimized procedures. Figure 1.8b shows the complete model-based electrical scheme: the first elaboration process adapts the $\overline{\mathbf{P}}$ parameters on the basis of the instantaneous environmental conditions; the second user-defined block solves, for each simulation step, the implicit equation: $\Gamma(I_{pv}, V_{pv}, \overline{\mathbf{P}}) = 0$, derived from (1.3). The solution of the implicit equation gives the set point to the controlled current generator that provides the current at devices connected to the PV field.

Of course, if iterative procedures are used for solving the embedded blocks, the computational burden can increase significantly, making the simulation too demanding. The approaches described in the following have been selected from the literature and represent a good trade-off in terms of accuracy and elaboration speed.

1.5.1 The Single-diode Model based on the Lambert W-function

Equation (1.3) can be expressed in explicit form if the Lambert W-function is used to calculate the solution of the exponential equation; if $y = xe^x$, then $x = W(y)$. Many details and useful references about the Lambert W-function can be found in the literature [10, 11].

After some manipulations, the PV current can be expressed as function of V_{pv} and $\overline{\mathbf{P}}$ as follows:

$$I_{pv} = F\left(V_{pv}, \overline{P}\right) = \frac{R_h \cdot \left(I_{ph} + I_s\right) - V_{pv}}{R_s + R_h} - \frac{\eta \cdot V_t}{R_s} \cdot W\left(\theta_I\right) \tag{1.14}$$

where:

$$\theta_I = \frac{\left(R_h // R_s\right) \cdot I_s \cdot e^{\frac{R_h \cdot R_s \cdot \left(I_{ph} + I_s\right) + R_h \cdot V_{pv}}{\eta \cdot V_t \cdot (R_h + R_s)}}}{\eta \cdot V_t} \tag{1.15}$$

with $x // y = x \cdot y/(x + y)$.

Alternatively the PV voltage can be expressed as function of I_{pv} and $\overline{\mathbf{P}}$ as follows:

$$V_{pv} = H\left(I_{pv}, \overline{P}\right) = R_h(I_{ph} + I_s) - (R_h + R_s)I_{pv} - \eta \cdot V_t \cdot W\left(\theta_V\right) \tag{1.16}$$

where:

$$\theta_V = \frac{R_h \cdot I_s \cdot e^{\frac{R_h \cdot \left(I_{ph} + I_s - I_{pv}\right)}{\eta \cdot V_t}}}{\eta \cdot V_t} \tag{1.17}$$

The explicit equations allow the easy implementation of the PV model in any simulation environment because no iterative procedure is required to find the operating point of the PV source. The only significant effort is in the calculation of $W(\theta)$. The Lambert W-function can be decomposed into branches expressed by simpler functions, so

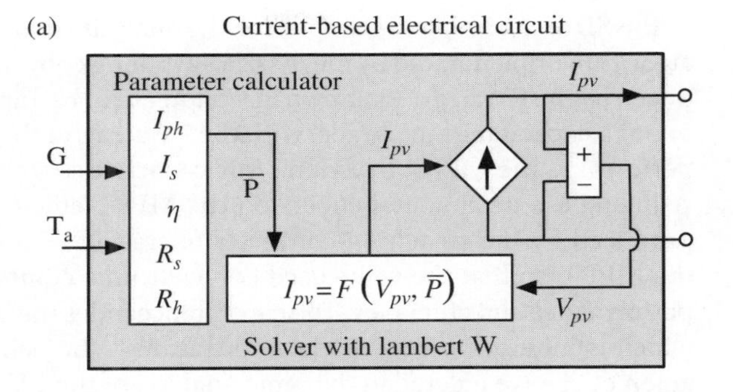

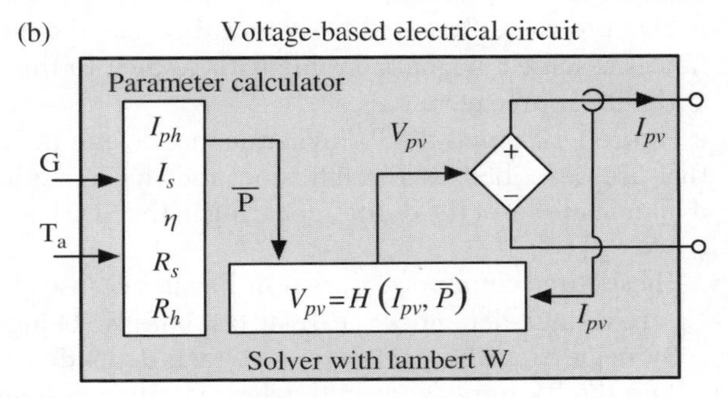

Figure 1.9 PV electrical schemes based on the Lambert W-function.

it can be implemented in any programming language [10, 11]. The schemes of Figure 1.9 show the simple blocks that might be implemented in the circuit simulator to obtain a lightweight, flexible, and reliable PV model by means of (1.14) and (1.16).

Batzelis et al. derived some approximate equations based on the Lambert W-function for calculating, in closed form, V_{MPP} and I_{MPP} too [12]:

$$V_{MPP} \simeq \frac{R_s + R_h}{R_h} \eta V_t (w-1) - R_s I_{ph} \left(1 - \frac{1}{w}\right) \tag{1.18}$$

$$I_{MPP} \simeq I_{ph} \left(1 - \frac{1}{w}\right) - \frac{\eta V_t (w-1)}{R_h} \tag{1.19}$$

where $w = W(I_{ph} e / I_s)$.

Unfortunately in non-uniform operating conditions, the PV models are not scalable and a different approach must be adopted. The extension of the methodologies shown in this chapter to non-uniform operating conditions will be outlined in Chapters 5 and 6.

1.6 PV Dynamic Models

The stationary models presented in Section 1.4, and especially the SDM, are widely adopted because they describe, by means of few components, all the main features of the PV device. They are able to reproduce with a high degree of accuracy the PV array non-linearity and the non-zero and the non-infinite slope of the I–V curve in the current and voltage generator regions, respectively. Moreover, as will be shown in Chapter 3, by adopting a suitable set of translational equations, it is also possible to simulate the PV source behavior at the desired values of irradiance and temperature.

The SDM is used, for example, for the analysis of the dynamic behavior of any MPPT algorithm implemented by means of a switching converter that processes the power produced by the PV array. As shown in Chapter 7, in real applications, it is common practice to put a capacitance in parallel with the PV array, at the input of the switching converter performing the MPPT function. This capacitance stabilizes the PV operating point by reducing the detrimental effects on the MPPT efficiency of the high-frequency ripple produced by the switching-converter operation. Unfortunately, the capacitance slows the MPPT, so that the value used comes from a compromise between MPPT dynamic performance and efficiency. The capacitance hides the dynamic behavior of the PV array, which is inherent in any diode, and thus also the behaviour of the PV source. Nousiainen et al. give a detailed dynamic analysis of the PV source [13], but major emphasis is given to the differential resistance, because the transition and the diffusion capacitances assume a negligible value with respect to the additional capacitance connected at the PV source terminals.

Figure 1.10 shows the PV dynamic model, and the two capacitive effects, C_D and C_T, that are described by the diffusion and the transition capacitances respectively. The dynamic model of the diode appearing in the SDM is completed by the differential resistance R_d [14].

These three components are non-linear, because all the three parameters R_d, C_T, and C_D are voltage-dependent [15], the last one also being dependent on the frequency [16].

By neglecting the parameters' voltage dependency, a linear model can be used to obtain the PV impedance in the desired range of frequencies. Simple algebra allows us to demonstrate that both the real and the imaginary part of the PV array impedance are frequency-dependent, not only explicitly but also through C_D. Indeed, by defining the equivalent parallel capacitance $C_p = C_T + C_D$ and resistance $R_p = R_h // R_d = R_h \cdot R_d / (R_h + R_d)$, the impedance is:

$$Z_{pv}(\omega) = R_s + R_p // \left(-j\frac{1}{\omega C_p}\right) = \left[R_s + \frac{R_p}{1 + (\omega R_p C_p)^2}\right] - j\frac{\omega R_p^2 C_p}{1 + (\omega R_p C_p)^2} \tag{1.20}$$

The impedance plot, in a Cartesian plane with the real and the imaginary parts of the impedance on the axes, is the best way of analyzing the frequency behavior of the PV source. Experimental results presented in the literature reveal that the typical impedance plot of the PV source is a semicircle. At a high frequency, the capacitances bypass the shunt resistance and the diode differential resistance, so that the semicircle intercepts the horizontal axis – the real part of the impedance – at a value equal to R_s.

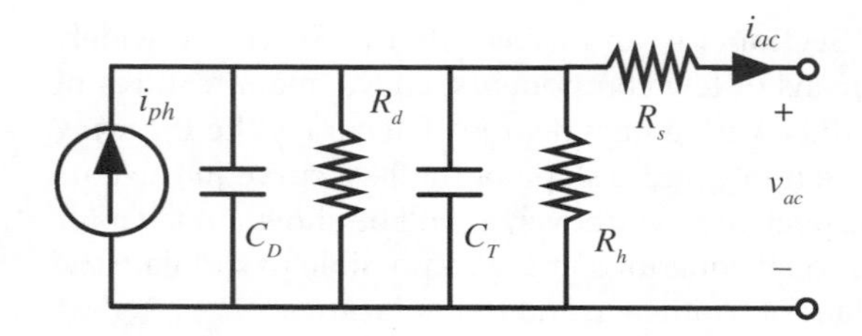

Figure 1.10 Dynamic model of the PV generator.

Meanwhile, at a very low frequency value, the capacitances behave as open circuits, so that the impedance plot intercepts the real axis at a value $R_s + R_p$. As shown by Johnson et al. for some types of PV modules [14], with cells produced with different cell technologies, this high-frequency value reaches a few hundreds of kilohertz. An example of this plot is given in Figure 1.11.

Having identified the values R_s and R_p by letting the PV source work at very high and at very low frequencies, the capacitance value C_p can be identified by analyzing the frequency behavior of the impedance. The value of C_p that minimizes the difference between the calculated impedance and the measured one is chosen [14].

This analysis can be done at any working condition of the PV generator, but values obtained in the dark are always indicated as the best in the literature. Indeed, in order to obtain the impedance plot experimentally, the PV generator must be stimulated by a sinusoidal signal (250 mV was used by Johnson et al. [14]) at different frequencies, so that so-called "impedance spectroscopy" is performed. The ratio between the phasors of the voltage and current signals at the frequency of the stimulus gives the impedance value at that frequency. The analysis in dark conditions allows avoidance of the injection of a current stimulus that must be kept considerably smaller than the average current value. The latter is the photo-induced current appearing during the day, which shows an almost linear dependency on the irradiance level.

As discussed by Chenvidhya et al. [17], in the dark, the relationship between the capacitive and the resistive effects changes significantly as a function of the PV cell's reverse or forward voltage bias. Parameters R_d, C_T, and C_D are significantly affected by the voltage bias, but also by the irradiation and temperature at which the PV generator works and by the cell's technology [18].

The capacitance remains quite low on the left-hand side of the maximum power point, the cell thus almost behaving as a current generator, but it increases dramatically when the open-circuit voltage is approached. A commonly used dependency of the capacitive effects on the operating voltage and the frequency is the following:

$$C_p(V,f) = \frac{k_1}{\left(1 - \frac{V}{k_2}\right)^{0.5}} + \frac{k_3}{2\pi f} e^{\frac{V}{k_4}} \left(\sqrt{1 + k_5 (2\pi f)^2} - 1\right)^{0.5} \tag{1.21}$$

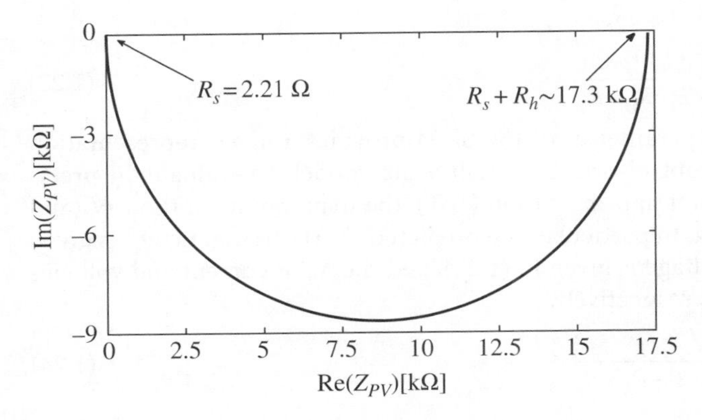

Figure 1.11 PV frequency behavior in the impedance plane; $R_p = 17.3\ \text{k}\Omega$, $R_s = 2.21\Omega$, $C_p = 2.2\mu\text{F}$.

An extensive analysis of such dependencies has recently appeared [19]. This experimental study is especially focused on frequency analysis of a PV string during its normal operation, both in the dark and at 1000 W/m^2 conditions, and in the presence of an hot spot affecting one cell. It has become evident that both the DC impedance and the equivalent capacitance increase when an hot spot occurs. The former can be measured at few tens of hertz and the latter at some tens of kilohertz, this value being dependent on the number of cells in the PV string. The string voltage at which such measurements are made has an effect that is significantly greater than the irradiance level. The model also takes into account the parasitic series inductance of the PV string [19].

1.7 PV Small-signal Models and Dynamic-resistance Modelling

Despite the precision and the widespread adoption of the SDM, it is difficult to include its non-linear and implicit equation in control-oriented models. Moreover, to design linear and some non-linear controllers it is necessary to linearize the system model around a given operating point. That small-signal model is available in automatic tools, such as the Simulink® linearization from MATLAB®, but such a mathematic linearization can lead to small-signal models of the PV array having incorrect behavior. In the following, some small-signal approximations that have been widely adopted for PV modules are analyzed.

The first attempt to obtain a small-signal model is to expand the exponential term of (1.3) into a Taylor series, as shown in (1.22). Since the analyses performed in small-signal models around a given operating point are usually based on Laplace representations, the adopted small-signal representation of the module is linear. This constrains the Taylor series to the first two terms, which leads to the small-signal model in (1.23), where $\hat{i}_{pv_{1oT}}$ and $\hat{v}_{pv_{1oT}}$ are small-signal quantities.

$$e^{B\cdot(V_{pv}+I_{pv}\cdot R_s)} \approx 1 + B\cdot(V_{pv}+I_{pv}\cdot R_s) \tag{1.22}$$

where, $B = \frac{1}{\eta V_t}$.

$$\hat{i}_{pv_{1oT}} = \frac{R_h\cdot I_{ph}}{R_h+R_s+I_s\cdot B\cdot R_h\cdot R_s} - \frac{(I_s\cdot B\cdot R_h+1)\cdot \hat{v}_{pv_{1oT}}}{R_h+R_s+I_s\cdot B\cdot R_h\cdot R_s} \tag{1.23}$$

Evaluating (1.23) with the parameters of the SDM provides a linear representation simple enough to develop a control-oriented small-signal model. To evaluate the precision of such a first-order Taylor approximation (1oT), the main points of the I–V (and P–V) curve must be analyzed. In particular, the predicted short-circuit current is given in (1.24), the open-circuit voltage is given in (1.25), and the MPP current and voltages are given in (1.26) and (1.27), respectively.

$$I_{sc_{1oT}} = \frac{R_h\cdot I_{ph}}{R_h+R_s+I_s\cdot B\cdot R_h\cdot R_s} \tag{1.24}$$

$$V_{oc_{1oT}} = \frac{R_h\cdot I_{ph}}{1+I_s\cdot B\cdot R_h} \tag{1.25}$$

$$I_{MPP_{1oT}} = \frac{1}{2} \cdot \frac{R_h \cdot I_{ph}}{R_h + R_s + I_s \cdot B \cdot R_h \cdot R_s} \tag{1.26}$$

$$V_{MPP_{1oT}} = \frac{1}{2} \cdot \frac{R_h \cdot I_{ph}}{1 + I_s \cdot B \cdot R_h} \tag{1.27}$$

Figure 1.12 shows a comparison between the SDM and the 1oT approximation. The simulation shows that 1oT model accurately represents the SDM from the short-circuit current up to near the MPP, only at its left-hand side. Unfortunately, the 1oT model exhibits large errors in both the MPP current and voltage and in the prediction of the open-circuit voltage. Taking into account that the PV module operates around the MPP most of the time due to the MPPT action [20–23], this model is not suitable for designing and analyzing control systems; the MPP of the 1oT approximation is very different from the MPP of the SDM model.

However, the structure of (1.23) can be used to design electrical models representing the PV module at the MPP, where two options are addressed: a current-based circuit and a voltage-based circuit. In the first case the PV current depends on the PV voltage as in (1.28), where the term I_{sc_N} corresponds to the model short-circuit current, while the slope stands for the model admittance. This is the Norton model and its electrical scheme is presented in Figure 1.13.

$$\hat{i}_{pv_N} = I_{sc_N} - \frac{1}{R_{pv_N}} \cdot \hat{v}_{pv_N} \tag{1.28}$$

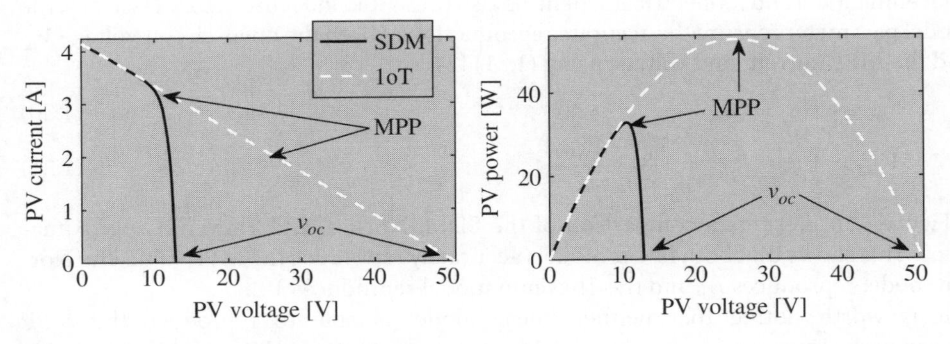

Figure 1.12 Comparison of SDM and 1oT models.

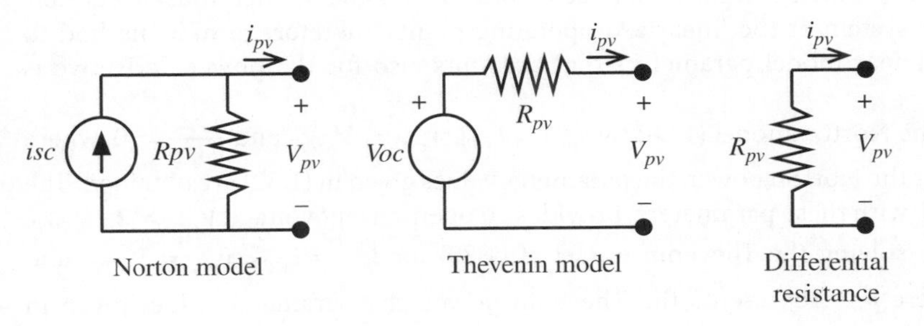

Figure 1.13 Linear small-signal models.

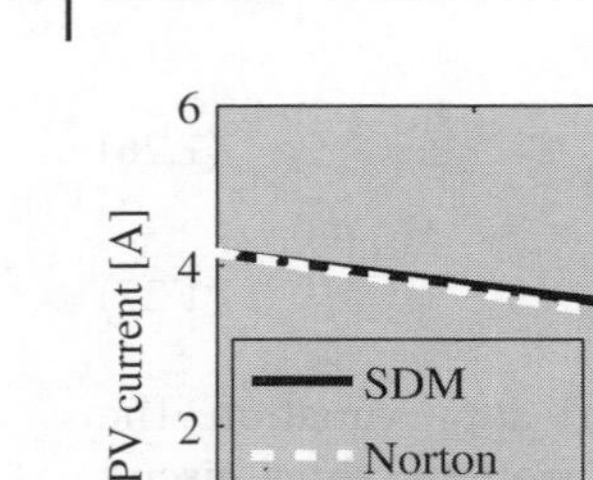

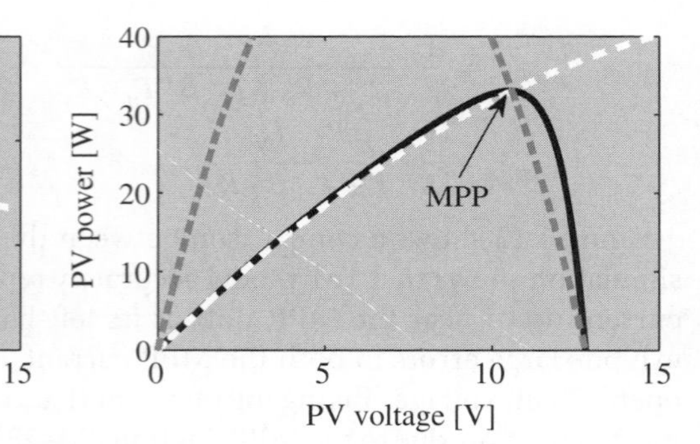

Figure 1.14 Comparison of the SDM, Norton, and Thevenin models.

The parameters I_{sc_N} and R_{pv_N} of this linear model can be calculated in different ways. A classical approach involves forcing the model to accurately reproduce both the short-circuit current I_{sc}, the MPP current I_{MPP}, and the voltage V_{MPP} using (1.29) [24].

$$I_{sc_N} = I_{sc}, \quad R_{pv_N} = \frac{V_{MPP}}{I_{sc} - I_{MPP}} \tag{1.29}$$

In the second approach – a voltage-based circuit – the PV voltage depends on the PV current as in (1.30), where the independent term V_{oc_T} corresponds to the model open-circuit voltage, while the slope stands for the model impedance R_{pv_T}. This is the Thevenin model and its electrical scheme is also presented in Figure 1.13. Classically, the model parameters ensure the accurate reproduction of both the open-circuit voltage V_{oc} and the MPP current and voltage using (1.31) [24].

$$\hat{v}_{pv_T} = V_{oc_T} - R_{pv_T} \cdot \hat{i}_{pv_T} \tag{1.30}$$

$$V_{oc_T} = V_{oc}, \quad R_{pv_T} = \frac{V_{oc} - V_{MPP}}{I_{MPP}} \tag{1.31}$$

Figure 1.14 presents a comparison of the SDM, Norton, and Thevenin approximations. This shows that both linear models accurately reproduce the MPP value, the Norton model reproduces I_{sc}, and the Thevenin model reproduces V_{oc}.

It is worth noting that neither linear model is able to reproduce the MPP power-derivative condition. Figure 1.14 shows that the MPP exhibits a power derivative equal to zero, while the Norton model predicts a positive derivative and the Thevenin model predicts a negative derivative, but small-signal models must accurately reproduce the system at the linearized operating point. Therefore, a new method to calculate the linear model parameters that accounts also for the power derivative is needed.

By solving the Norton model (1.28) for $\hat{i}_{pv_N} = I_{MPP}$, $\hat{v}_{pv_N} = V_{MPP}$, and $\frac{d\,p_{pv_N}}{d\,v_{pv_N}} = 0$, where p_{pv_N} represents the Norton power, the parameter values given in (1.32) are obtained. The Norton model, with these parameters, provides an open-circuit voltage $V_{oc_N} = 2 \cdot V_{MPP}$.

Similarly, by solving the Thevenin model of (1.30) for $\hat{i}_{pv_T} = I_{MPP}$, $\hat{v}_{pv_T} = V_{MPP}$, and $\frac{d\,p_{pv_T}}{d\,v_{pv_T}} = 0$, where p_{pv_T} represents the Thevenin power, the parameter values given in (1.33) are obtained.

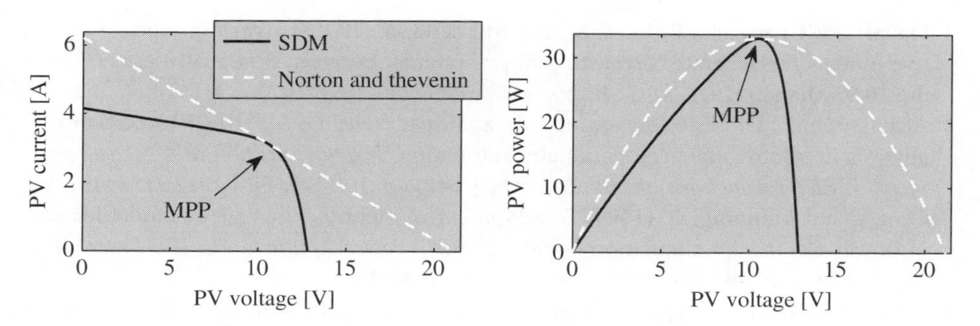

Figure 1.15 Comparison of SMD, Norton, and Thevenin models with accurate MPP approximation.

$$I_{sc_N} = 2 \cdot I_{MPP}, \quad R_{pv_N} = \frac{V_{MPP}}{I_{MPP}} \tag{1.32}$$

$$V_{oc_T} = 2 \cdot V_{MPP}, \quad R_{oc_T} = \frac{V_{MPP}}{I_{MPP}} \tag{1.33}$$

Recalling the linear circuit theory, (1.32) and (1.33) reveal that both Norton and Thevenin circuits parameterized at the MPP are equivalent, as expected. Figure 1.15 shows a comparison between the SDM and the linear small-signal models' reproduction of MPP: it can be seen that both the value and the derivative of the the power at the MPP are accurately provided by the linear model. Hence the Norton or Thevenin approximations, parameterized as in (1.32) and (1.33) respectively, are good candidates for developing a linearized model of the complete PV system, including a model of the power processing system, for control purposes.

Finally, a simpler model, named the differential resistance model, is also found in the literature [20, 21]. This model, also depicted in Figure 1.13, consists of a single impedance representing the PV module. The differential resistance magnitude is obtained from both the Norton and Thevenin models, parameterized as in (1.32) and (1.33) respectively, by nullifying the corresponding source: opening the current source in the Norton model or short-circuiting the voltage source in the Thevenin model. As expected, the differential resistance is negative, since it is derived by adopting the generator convention between the current and voltage references at the generator terminals. The differential resistance is useful for analyzing the behavior of power converters at the MPP, removing the input added to the linear system by the source of the complete linear models. Unfortunately, this characteristic shows a limitation: it is not possible to analyze the behavior of the PV system in the presence of irradiance (represented by the current source of the Norton model) or temperature (represented by the voltage source of the Thevenin model) variations.

References

1 Syafaruddin, Karatepe, E., and Hiyama, T. (2009) Artificial neural network-polar coordinated fuzzy controller based maximum power point tracking control under partially shaded conditions. *Renewable Power Generation, IET*, **3** (2), 239 –253, doi:10.1049/iet-rpg:20080065.

2 Ogliari, E., Grimaccia, F., Leva, S., and Mussetta, M. (2013) Hybrid predictive models for accurate forecasting in pv systems. *Energies*, **6** (4), 1918–1929, doi:10.3390/en6041918. URL: http://www.mdpi.com/1996-1073/6/4/1918.

3 Attivissimo, F., Di Nisio, A., Savino, M., and Spadavecchia, M. (2012) Uncertainty analysis in photovoltaic cell parameter estimation. *Instrumentation and Measurement, IEEE Transactions on*, **61** (5), 1334–1342, doi:10.1109/TIM.2012.2183429.

4 Gow, J. and Manning, C. (1999) Development of a photovoltaic array model for use in power-electronics simulation studies. *Electric Power Applications, IEE Proceedings -*, **146** (2), 193–200, doi:10.1049/ip-epa:19990116.

5 Khanna, V., Das, B., Bisht, D., Vandana, and Singh, P. (2015) A three diode model for industrial solar cells and estimation of solar cell parameters using {PSO} algorithm. *Renewable Energy*, **78**, 105–113, doi:10.1016/j.renene.2014.12.072. URL: http://www.sciencedirect.com/science/article/pii/S0960148115000063.

6 Hejri, M., Mokhtari, H., Azizian, M., Ghandhari, M., and Soder, L. (2014) On the parameter extraction of a five-parameter double-diode model of photovoltaic cells and modules. *Photovoltaics, IEEE Journal of*, **4** (3), 915–923, doi:10.1109/JPHOTOV.2014.2307161.

7 Babu, B. and Gurjar, S. (2014) A novel simplified two-diode model of photovoltaic (pv) module. *Photovoltaics, IEEE Journal of*, **4** (4), 1156–1161, doi:10.1109/JPHOTOV.2014.2316371.

8 Muhsen, D.H., Ghazali, A.B., Khatib, T., and Abed, I.A. (2015) Parameters extraction of double diode photovoltaic module's model based on hybrid evolutionary algorithm. *Energy Conversion and Management*, **105**, 552–561, doi:10.1016/j.enconman.2015.08.023. URL: http://www.sciencedirect.com/science/article/pii/S0196890415007694.

9 Miceli, R., Orioli, A., and Gangi, A.D. (2015) A procedure to calculate the i-v characteristics of thin-film photovoltaic modules using an explicit rational form. *Applied Energy*, **155**, 613–628, doi:10.1016/j.apenergy.2015.06.037. URL: http://www.sciencedirect.com/science/article/pii/S0306261915007953.

10 Weisstein, E. (2008) Lambert w-function. *A Wolfram Web Resource. http://mathworld.wolfram.com/LambertW-Function.html.*

11 Corless, R., Gonnet, G., Hare, D., Jeffrey, D., and Knuth, D. (1996) On the lambertw function. *Advances in Computational Mathematics*, **5** (1), 329–359, doi:10.1007/BF02124750.

12 Batzelis, E., Kampitsis, G., Papathanassiou, S., and Manias, S. (2015) Direct mpp calculation in terms of the single-diode pv model parameters. *Energy Conversion, IEEE Transactions on*, **30** (1), 226–236, doi:10.1109/TEC.2014.2356017.

13 Nousiainen, L., Puukko, J., Mäki, A., Messo, T., Huusari, J., Jokipii, J., Viinamäki, J., Lobera, D., Valkealahti, S., and Suntio, T. (2013) Photovoltaic generator as an input source for power electronic converters. *Power Electronics, IEEE Transactions on*, **28** (6), 3028–3038, doi:10.1109/TPEL.2012.2209899.

14 Johnson, J., Schoenwald, D., Kuszmaul, S., and Strauch, J. (2011) Creating dynamic equivalent pv circuit models with impedance spectroscopy for arc-fault modeling. *SANDIA Report*, **SAND2011-4247**.

15 Kumar, R., Suresh, M.S., and Nagaraju, J. (2006) Effect of solar array capacitance on the performance of switching shunt voltage regulator. *Power Electronics, IEEE Transactions on*, **21** (2), 543–548, doi:10.1109/TPEL.2005.869779.

16 Thongpron, J., Kirtikara, K., and Jivacate, C. (2006) A method for the determination of dynamic resistance of photovoltaic modules under illumination. *Solar Energy Materials and Solar Cells*, **90** (18 19), 3078–3084, doi:10.1016/j.solmat.2006.06.029. URL: http://www.sciencedirect.com/science/article/pii/S0927024806002200, 14th International Photovoltaic Science and Engineering Conference 14th International Photovoltaic Science and Engineering Conference.

17 Chenvidhya, D., Kirtikara, K., and Jivacate, C. (2005) {PV} module dynamic impedance and its voltage and frequency dependencies. *Solar Energy Materials and Solar Cells*, **86** (2), 243–251, doi:10.1016/j.solmat.2004.07.005. URL: http://www.sciencedirect.com/science/article/pii/S0927024804002910.

18 Kumar, R.A., Suresh, M., and Nagaraju, J. (2004) Gaas/ge solar cell {AC} parameters under illumination. *Solar Energy*, **76** (4), 417–421, doi:10.1016/j.solener.2003.10.004. URL: http://www.sciencedirect.com/science/article/pii/S0038092X03004237.

19 Kim, K., Seo, G.S., Cho, B.H., and Krein, P. (2015) Photovoltaic hot spot detection for solar panel substrings using ac parameter characterization. *Power Electronics, IEEE Transactions on*, **PP** (99), 1–1, doi:10.1109/TPEL.2015.2417548.

20 Femia, N., Petrone, G., Spagnuolo, G., and Vitelli, M. (2005) Optimization of perturb and observe maximum power point tracking method. *IEEE Transactions on Power Electronics*, **20** (4), 963–973.

21 Femia, N., Petrone, G., Spagnuolo, G., and Vitelli, M. (2009) A technique for improving p&o mppt performances of double-stage grid-connected photovoltaic systems. *IEEE Transactions on Industrial Electronics*, **56** (11), 4473–4482.

22 Bianconi, E., Calvente, J., Giral, R., Mamarelis, E., Petrone, G., Ramos-Paja, C.A., Spagnuolo, G., and Vitelli, M. (2013) Perturb and observe mppt algorithm with a current controller based on the sliding mode. *International Journal of Electrical Power & Energy Systems*, **44** (1), 346–356.

23 Bianconi, E., Calvente, J., Giral, R., Mamarelis, E., Petrone, G., Ramos-Paja, C., Spagnuolo, G., and Vitelli, M. (2013) A fast current-based mppt technique employing sliding mode control. *IEEE Transactions on Industrial Electronics*, **60** (3), 1168–1178.

24 Trejos, A., Gonzalez, D., and Ramos-Paja, C.A. (2012) Modeling of step-up grid-connected photovoltaic systems for control purposes. *Energies*, **5** (6), 1900–1926.

2

Single-diode Model Parameter Identification

2.1 Introduction

The capability of a PV model to accurately reproduce the I–V curve of a photovoltaic source does not depend exclusively on the model complexity; it also depends on the parameter values appearing in the model, which are adapted so as to reproduce the behavior of different PV cells, panels, and arrays. Thus an important question in PV model construction is the method used for calculating these parameters. As already shown in Chapter 1, in some cases, simplified models are adopted, because their parameters can be readily extracted from the datasheets provided by PV manufacturers or from experimental data. On the other hand, complex models give a more precise representation and so, where possible, they are preferred. The SDM is a good compromise between precision and complexity, so this chapter is mainly focused on the identification of the values of the five parameters $\overline{\mathbf{P}} = [I_{ph}, I_s, \eta, R_s, R_h]$ in this model for given working conditions. The procedures described in this chapter refer to the calculation of $\overline{\mathbf{P}}$ for the PV panel model, but under the assumption of uniform conditions the SDM parameters can be scaled up or down, as discussed in Chapter 1.

Unfortunately, no standardized method has been defined by the scientific community for evaluating $\overline{\mathbf{P}}$. Nevertheless, many techniques have been proposed in the literature for obtaining the values of the five parameters: using information provided by the PV manufacturers, by performing some measurements in the real world, or both. In general, parameter identification requires the solution of a system of non-linear equations and different numerical methods have been used for calculating its exact solution. A complementary strategy is adopted in approximated methods where, under certain reasonable simplifications, the parameters are calculated by means of explicit equations. In this second case the reduced accuracy is compensated for by a lower computational effort. In this chapter, some of the main methods presented in literature are described and their applications in real cases are proposed.

2.2 PV Parameter Identification from Datasheet Information

2.2.1 Exact Numerical Methods

Given the data that are made available by the manufacturer for a PV module, a system of non-linear equations is constructed by forcing (1.3) to fit the I–V curve at various

Photovoltaic Sources Modeling, First Edition. Giovanni Petrone, Carlos Andrés Ramos-Paja and Giovanni Spagnuolo.

Companion Website: www.wiley.com/go/petrone/Photovoltaic_Sources_Modeling

operating points in order to have a sufficient number of equations to calculate the values of I_{ph}, I_s, η, R_s, and R_h. The short-circuit, maximum power, and open-circuit points are the most common choice because of their easy identification on the I–V curve and because they are always provided by the PV manufacturer in the datasheet, at least in STC:

$$F_1 : I_{ph} - I_s\left(e^{\left(\frac{R_s \cdot I_{SC}}{\eta \cdot Vt}\right)} - 1\right) - \frac{R_s \cdot I_{sc}}{R_h} - I_{sc} = 0 \tag{2.1}$$

$$F_2 : I_{ph} - I_s\left(e^{\left(\frac{V_{MPP}+R_s \cdot I_{MPP}}{\eta \cdot Vt}\right)} - 1\right) + \\ - \frac{V_{MPP} + R_s \cdot I_{MPP}}{R_h} - I_{MPP} = 0 \tag{2.2}$$

$$F_3 : I_{ph} - I_s\left(e^{\left(\frac{V_{oc}}{\eta \cdot Vt}\right)} - 1\right) - \frac{V_{oc}}{R_h} = 0 \tag{2.3}$$

Further non-linear equations result from the derivative of the power and the derivative of the current with respect to the voltage at the MPP:

$$F_4 : \left.\frac{\partial P_{pv}}{\partial V_{pv}}\right|_{MPP} = 0 \tag{2.4}$$

$$F_5 : \left.\frac{\partial I_{pv}}{\partial V_{pv}}\right|_{MPP} = -\frac{I_{MPP}}{V_{MPP}} \tag{2.5}$$

It is worth noting that (2.5) and (2.4), although related as in (2.6), give rise to two independent non-linear equations:

$$\frac{\partial P_{pv}}{\partial V_{pv}} = \left.\frac{\partial\left(V_{pv} \cdot I_{pv}\right)}{\partial V_{pv}}\right|_{MPP} = 0 \rightarrow I_{MPP} + V_{MPP} \cdot \left.\frac{\partial I_{pv}}{\partial V_{pv}}\right|_{MPP} = 0 \tag{2.6}$$

The derivative of the current with respect to the voltage is calculated from (1.3) and has the following expression:

$$\left.\frac{\partial I_{pv}}{\partial V_{pv}}\right|_{MPP} = -\frac{\frac{1}{R_h} + \frac{I_s}{\eta \cdot V_t} \cdot e^{\frac{V_{MPP}+I_{MPP} \cdot R_s}{\eta V_t}}}{1 + \frac{R_s}{R_h} + \frac{Rs \cdot I_s}{\eta \cdot V_t} \cdot e^{\frac{V_{MPP}+I_{MPP} \cdot R_s}{\eta V_t}}} \tag{2.7}$$

The derivative of the power with respect to the voltage is given by:

$$\left.\frac{\partial P_{pv}}{\partial V_{pv}}\right|_{MPP} = \frac{\eta V_t[(I_{ph} + I_s)R_h - 2V_{MPP}]}{I_s R_s R_h e^{\frac{I_{MPP}R_s+V_{MPP}}{\eta V_t}} + \eta V_t(R_s + R_h)} \\ + \frac{I_s R_h e^{\frac{I_{MPP}R_s+V_{MPP}}{\eta V_t}}\left[I_{MPP}R_s - (V_{MPP} + \eta V_t)\right]}{I_s R_s R_h e^{\frac{I_{MPP}R_s+V_{MPP}}{\eta V_t}} + \eta V_t(R_s + R_h)} \tag{2.8}$$

and it is obtained by deriving the following PV power expression:

$$P_{pv} = V_{pv} \cdot I_{pv} = V_{pv} \cdot I_{ph} - V_{pv} \cdot I_s\left(e^{\frac{V_{pv}^2+R_s \cdot P_{pv}}{V_{pv}\eta Vt}} - 1\right) - \frac{V_{pv}^2 + R_s \cdot P_{pv}}{R_h} \tag{2.9}$$

The system of non-linear equations formed by (2.1)–(2.5) can be solved numerically using iterative procedures based on curve-fitting or root-finding. The bisection method and the Newton–Raphson method, belonging to the root-finding category, are the most familiar procedures for finding the solution of the non-linear system. The curve-fitting methods, such as the Levenberg–Marquardt algorithm, are also applicable, but they are much more effective when a large number of experimental points are available to describe the I–V curve. Although such approaches provide precise results in some cases, especially the Newton–Raphson method, the result is sensitive to the initial guess solution and convergence might not be reached at all. In order to overcome such drawbacks many alternative methods have been developed and optimized for achieving reliable solutions.

Lo Brano and Ciulla formalized the solution of the non-linear system formed by (2.1)–(2.5) as an optimization problem in which the following function must be minimized [1]:

$$\Upsilon(I_{ph}, I_s, \eta, R_s, R_h) = \sum_{i=1}^{5} F_i^2 \tag{2.10}$$

The generalized reduced gradient algorithm was applied to solve (2.10).

The five parameter values have also been found using genetic algorithms [2] and neural networks [3].

In order to reduce the complexity of the non-linear system, alternative equations can be used. For example, instead of (2.4), some authors prefer to use the relation between the open-circuit voltage and the temperature coefficient α_v by imposing the following condition:

$$\frac{\partial V_{oc}}{\partial T} = \alpha_v \tag{2.11}$$

Tian et al. solved the system of non-linear equations including (2.1)–(2.3), (2.5), and (2.11) in MATLAB® using *fsolve* and by considering the PV panel operating in STC [4].

Other authors have suggested introducing some approximations to the non-linear equations in order to obtain more simple or explicit expressions for evaluating the unknown parameters. Of course, in this case the simplicity comes at the cost of an approximate set of SDM parameters.

Chan and Phang treated the exponential term as a third-order polynomial expression by using the Taylor expansion [5]; the procedure has been applied for both SDM and DDM.

Other authors have obtained the five parameters in two steps [6, 7]: first they assume $I_{ph} \simeq I_{sc}$ and $R_h \to \infty$ and then, using (2.3) and (2.11), the values of I_s and η are evaluated explicitly. In a second phase, the assumption $R_h \to \infty$ is removed and the two non-linear equations (2.2) and (2.5) are solved iteratively to find R_s and R_h values.

Sera et al. [8] replace (2.5) with the approximated condition:

$$\left.\frac{\partial I}{\partial V}\right|_{I_{pv}=I_{sc}} \simeq -\frac{1}{R_{h0}} \tag{2.12}$$

and then, after making some simplifications, the system of five non-linear equations is split into a system of three non-linear equations to find η, R_s, and R_h, and into two independent equations for calculating I_{ph} and I_s.

Orioli and Di Gangi [9] calculated the five parameters using the non-linear equations (2.1)–(2.3) coupled with (2.12) and one referring to the open-circuit condition:

$$\left.\frac{\partial I}{\partial V}\right|_{V_{pv}=V_{oc}} \simeq -\frac{1}{R_{s0}} \tag{2.13}$$

By making some approximations, all the parameters can be expressed as functions of the series resistance and then, starting from an initial value for R_s, the values of the five parameters are obtained with a simple trial-and-error procedure. The effectiveness of such a simplified procedure has already been proven in the literature; it is among the most widely adopted procedures. The choice of the most appropriate method is therefore almost exclusively dependent on the computational resources required to solve the system of non-linear equations.

In the next section an alternative compact and efficient approach will be described in detail, and some examples will demonstrate its accuracy.

2.2.2 Approximate Explicit Solution for Calculating SDM Parameters

The procedure shown in this section was first introduced by Femia et al. [7] and Ramos-Paja et al. [10], who applied it to evaluate the parameters of the SDM in STC. Here their approach is extended to extract the SDM parameters for any environmental condition. The method exploits the Lambert W-function to obtain a set of explicit equations suitable for implementation in any computational software.

The assumption $R_h \to \infty$ allows the current in the R_h branch to be neglected, simplifying (1.3) [6]. A further simplification can be made in (2.1) because when the PV panel is in the short-circuit operating condition, the diode included in the model is reverse biased and its current is significantly lower than the photo-induced current I_{ph}. The following approximation is therefore valid for any environmental operating condition:

$$I_{ph} \simeq I_{sc} \tag{2.14}$$

Using (2.14), from (2.3) the ideality factor is expressed as:

$$\eta = \frac{V_{oc}}{V_t \cdot \ln\left(\frac{I_{sc}}{I_s}+1\right)} \tag{2.15}$$

As for the saturation current I_s, it is common practice to consider only its temperature variation [6, 7, 11, 12]; this is primarily due to the temperature dependence of intrinsic carrier generation and can be approximated as:

$$I_s = C_0 \cdot T^3 e^{\left(-\frac{E_g}{kT}\right)} \tag{2.16}$$

where C_0 is the temperature coefficient (A K^{-3}). Usually C_0 is obtained by a fitting procedure; the mathematical steps for calculating C_0 by using the datasheet information can be found in the literature [6, 7]. The final results of this procedure are the following equations:

$$C_0 = \frac{I_{sc0} \cdot e^{\gamma_0}}{T_0^3} \tag{2.17}$$

$$\gamma_0 = -\frac{V_{oc0}}{\alpha_V - \frac{V_{oc0}}{T_0}}\left(\frac{\alpha_I}{I_{sc0}} - \frac{3}{T_0} - \frac{E_{g0}}{kT_0^2}\right) + \frac{E_{g0}}{kT_0} \tag{2.18}$$

Table 2.1 Energy bandgap at $T_{ref} = 25°C$.

	Silicon thin film [J]	Single-crystalline [J]	Polycrystalline [J]	Three-junction amorphous [J]
E_g	$1.794 \cdot 10^{-19}$	$1.794 \cdot 10^{-19}$	$1.826 \cdot 10^{-19}$	$2.563 \cdot 10^{-19}$

The subscript "0" stands for a reference condition, usually corresponding to STC. E_g is the material bandgap; in Table 2.1 the values of E_g evaluated at 25°C are shown for different cell technologies.

Boyd et al. [13] introduced in the diode reverse-saturation current I_s a correction factor accounting for the dependence on the irradiance value which modifies (2.16) into:

$$I_s = C_0 \cdot T^3 \left(\frac{G_0}{G}\right)^m e^{\left(-\frac{E_g}{kT}\right)} \tag{2.19}$$

where the parameter m is determined by fitting the condition $\partial P_{pv}/\partial V_{pv} = 0$ for the MPP at 200 W/m² and 25°C. This fit assures better modeling results at low irradiance. PV panel manufacturers should provide the maximum-power current and voltage at 200 W/m² and 25°C, as this information is required by the California Energy Commission [14], but empirical values for m can be adopted too.[1]

Then, only R_s and R_h remain to be determined; the following change of variable is first introduced:

$$x = \frac{V_{MPP} + R_s \cdot I_{MPP}}{\eta V_t} \tag{2.20}$$

This allows the series resistance to be expressed as:

$$R_s = \frac{x\eta V_t - V_{MPP}}{I_{MPP}} \tag{2.21}$$

The parallel resistance, from (2.2), can be expressed as:

$$R_h = \frac{V_{MPP} + I_{MPP} \cdot R_s}{I_{ph} - I_{MPP} - I_s \cdot \left(e^{\frac{V_{MPP}+I_{MPP}\cdot R_s}{\eta V_t}} - 1\right)} \tag{2.22}$$

so that:

$$R_h = \frac{x\eta V_t}{I_{ph} - I_{MPP} - I_s \cdot (e^x - 1)} \tag{2.23}$$

The substitution of (2.20), (2.21), and (2.23) in (2.5) leads to:

$$\frac{I_{MPP}}{V_{MPP}} - \frac{1 + \frac{I_s x e^x}{I_{ph}-I_{MPP}-I_s(e^x-1)}}{\frac{e^x I_s(x^2\eta V_t - x V_{MPP})}{I_{MPP}\left[I_{ph}-I_{MPP}-I_s(e^x-1)\right]} + \frac{\eta x V_t}{I_{ph}-I_{MPP}-I_s(e^x-1)} + \frac{\eta x V_t - V_{MPP}}{I_{MPP}}} = 0 \tag{2.24}$$

1 For example, in the paper by Boyd et al. [13], $m = 0.278$ for crystalline silicon panels and $m = 1.34$ for two layers of amorphous silicon (2-a-Si).

In the denominator of (2.24) the term $x^2\eta V_t$ appears. By squaring (2.20) and by neglecting the term depending on the small quantity R_s^2, a few algebraic steps lead to:

$$x^2\eta V_t \approx -\frac{V_{MPP}^2}{\eta V_t} + 2 \cdot V_{MPP} \cdot x \tag{2.25}$$

Consequently, solving (2.24) means solving (2.26) with respect to x:

$$\begin{aligned} &2V_{MPP} \cdot (I_{MPP} - I_{ph} - I_s) + (I_{ph} + I_s)\, \eta V_t \cdot x \\ &+ I_s \cdot e^x \left[-\eta V_t x + V_{MPP} \cdot \left(2 - \frac{V_{MPP}}{\eta V_t} \right) \right] = 0 \end{aligned} \tag{2.26}$$

By accounting for (2.20) and by defining $I^* = I_{ph} + I_s$, the first two terms in (2.26) can be expressed and then simplified as follows:

$$\begin{aligned} &2V_{MPP} \cdot (I_{MPP} - I^*) + I^*\eta V_t x \\ &= V_{MPP} \cdot (2 \cdot I_{MPP} + I^*) + I^* \cdot (R_s \cdot I_{MPP} - 2 \cdot V_{MPP}) \\ &\approx V_{MPP} \cdot (2 \cdot I_{MPP} + I^*) - 2 \cdot I^* \cdot V_{MPP} = V_{MPP} \cdot (2 \cdot I_{MPP} - I^*) \end{aligned} \tag{2.27}$$

The approximation adopted in (2.27) is justified by the fact that, for a large class of PV modules on the market, the quantity $R_s \cdot I_{MPP}$ is almost two orders of magnitude smaller than $2 \cdot V_{MPP}$. The series resistance value as well as the current and voltage values at the MPP have been reported for a large number of commercial modules in the literature [9]. Looking at those numbers it is easy to verify that the assumption made above is reasonable. Thus (2.26) can be written as follows:

$$\frac{V_{MPP} \cdot (2 \cdot I_{MPP} - I_{ph})}{\eta V_t I_s} - e^x \left[x - \left(\frac{2V_{MPP}}{\eta V_t} - \frac{V_{MPP}^2}{\eta^2 V_t^2} \right) \right] = 0 \tag{2.28}$$

and by performing the variable change:

$$\sigma = x - \left(\frac{2V_{MPP}}{\eta V_t} - \frac{V_{mpp}^2}{\eta^2 V_t^2} \right) \tag{2.29}$$

Equation 2.28 is equivalent to:

$$\frac{V_{MPP} \cdot (2 \cdot I_{MPP} - I_{ph})\, e^{\frac{V_{MPP}(V_{MPP} - 2\eta V_t)}{\eta^2 {V_t}^2}}}{\eta V_t I_s} - e^{\sigma}\sigma = 0 \tag{2.30}$$

As noted in Section 1.5.1, the solution of $\sigma e^{\sigma} = \delta$ is $\sigma = W(\delta)$, so the value of x is given by inverting (2.30) as follows:

$$x = W\left[\frac{V_{MPP}\,(2I_{MPP} - I_{ph})\, e^{\frac{V_{MPP}(V_{MPP} - 2\eta V_t)}{\eta^2 {V_t}^2}}}{\eta I_s V_t} \right] + 2\frac{V_{MPP}}{\eta V_t} - \frac{V_{MPP}^2}{\eta^2 {V_t}^2} \tag{2.31}$$

The proposed procedure does not require the solution of any non-linear system, thus requiring only the use of (2.14), (2.15), and (2.19) to calculate straightforwardly I_{ph}, η, and I_s; subsequently the intermediate variable x is calculated with (2.31) and then used in (2.21) and (2.23) to estimate the values of the parallel and series resistances.

As shown by the equations, the five parameters can be calculated only by knowing I_{sc}, V_{oc}, V_{MPP}, I_{MPP}, and the environmental parameters G and T. Usually such data can be measured or estimated in both indoor and outdoor conditions. For example, if it is of interest to reproduce the I–V curve at STC, the five parameters are calculated using only data taken from the module datasheet; the same set of simplified equations can be used to tune the SDM using experimental data for any arbitrary environmental condition, and assuring the same degree of accuracy. It is worth noting that the explicit formulas can also be used to calculate the guess solution of the root-fitting and curve-fitting methods, as described in Section 2.2.1, accelerating the convergence of those methods and leading them to a more accurate result.

2.2.3 Validation of the Approximate Explicit Solution

The equations obtained in the previous section are now validated, using as examples the Kyocera KC175GHT-2 and Sanyo HIT240HDE-4 commercial panels. The set of the estimated parameters (I_{ph}, I_s, η, R_s, R_h) will be compared with the set of parameters obtained by solving the system of non-linear equations in (2.1)–(2.3) by means of the generalized reduced gradient algorithm [1]. Initially, the analysis will be conducted by assuming the two panels are operating at STC; subsequently some comparisons at different irradiance conditions will be provided. Table 2.2 shows the datasheet information used for extracting the unknown parameters.

In Table 2.3 the five estimated parameters obtained [1] are reported in columns headed "Exact", since they are obtained by solving the system of exact equations. The values calculated with the approximate explicit equations, introduced in the previous section, are in the columns headed "Approximate". It is worth noting that the exact solution is achieved only if very high precision is adopted during the numerical process for solving the system of equations [1]. This requirement might be a strong limitation if

Table 2.2 Datasheet information at STC.

	V_{oc} [V]	I_{sc} [A]	V_{MPP} [V]	I_{MPP} [A]	α_V [mV/°C]	α_I [mA/°C]
Kyocera KC175GHT-2	29.2	8.09	23.6	7.42	−109	3.18
Sanyo HIT240HDE	43.6	7.37	35.5	6.77	−109	2.21

For $G = 1000\ \text{W/m}^2$ and $T = 25°\text{C}$.

Table 2.3 Values of the five parameters of Kyocera and Sanyo panels at STC.

	Kyocera KC175GHT-2		Sanyo HIT240HDE	
Solution	**Exact**	**Approximate**	**Exact**	**Approximate**
I_s [nA]	1.0660×10^{-1}	2.0722	8.2580×10^{-2}	1.7730×10^{-2}
I_{ph} [A]	8.1175	8.09	7.3925	7.37
R_s [Ω]	0.2836	0.2185	0.4249	0.4223
R_h [Ω]	83.3021	93.0571	139.2965	126.0704
η [/]	0.9474	1.0730	1.1244	1.0581

the method must be implemented on an embedded system dedicated to the monitoring and diagnosis of a PV plant, where the computational resources are quite limited because cheap microcontrollers or digital signal processors (DSP) are usually used for the numerical elaboration. In contrast, the only effort involved in evaluating the simplified equations is in the use of the Lambert W-function, which can be decomposed into branches expressed by simpler functions, so it is easy to represent in any programming language [15, 16]. Alternatively the Lambert W-function values can be loaded into a lookup table, thus avoiding any numerical evaluation on the embedded system.

The data shown in Table 2.3 highlight that the approximate solution gives values that are close to those of the exact solution, except for the saturation current I_s, where a very large difference appears. However, comparing the two solutions in terms of I–V and and P–V curves, the difference is absolutely acceptable. Indeed, as shown in Figure 2.1, the curves reconstructed with the approximate solution almost overlap those corresponding to the exact solution for both PV panels under study. In the same figure the white circles represent the short-circuit current, the MPP and the open-circuit voltage taken from the PV panel datasheet.

In order to quantify the differences, the error in the PV current has been calculated, point by point, as follows:

$$\mathbf{ERR_I}\% = \frac{I_{pv,Approx} - I_{pv,Exact}}{I_{sc,Exact}} \cdot 100 \quad \forall\, V_{pv} \in [0, V_{oc}] \tag{2.32}$$

It is worth noting that the error has been normalized with respect to the short-circuit current $I_{sc,Exact}$ in order to appreciate the error variations over the whole range of the PV voltage, from zero to the open-circuit voltage.

The results of (2.32) for the two panels under study are reported in Figure 2.2; these plots show that the error is almost confined to a band of ±0.5%, except for the region close to the open-circuit voltage. The white circles identify the MPPs, where the error tends to zero, showing the high precision of the simplified approach close to the MPPs. The largest error in the proximity of the open-circuit voltage is due to the higher slope of the I–V curves, so the current calculated with the two sets of the $\overline{\mathbf{P}}$ parameters is significantly different for the same value of the PV voltage. Close to the open-circuit voltage it is more appropriate to evaluate the voltage error instead of the current error, but this analysis is superfluous because the good match at the open-circuit voltage is already evident in Figure 2.1.

One of the main advantages of this procedure is the ability to also evaluate the parameters of the SDM when the environmental operating conditions change, provided that the short-circuit current, the open-circuit voltage, and the current and voltage at the MPP are known, or properly measured.

In order to verify the accuracy of the proposed explicit equations, further comparisons have been conducted on the Kyocera panel using the experimental I–V curves in different irradiance conditions. The experimental data have been extracted from the module datasheet and the five parameter values have been calculated for these cases. Figure 2.3 compares the experimental data and the model. The agreement is always acceptable, as confirmed by the percentage error of the current shown in Figure 2.4. The error is a little higher than the error in the comparison with the exact solution [1], because the experimental data reported in the datasheet were in graphical form and a

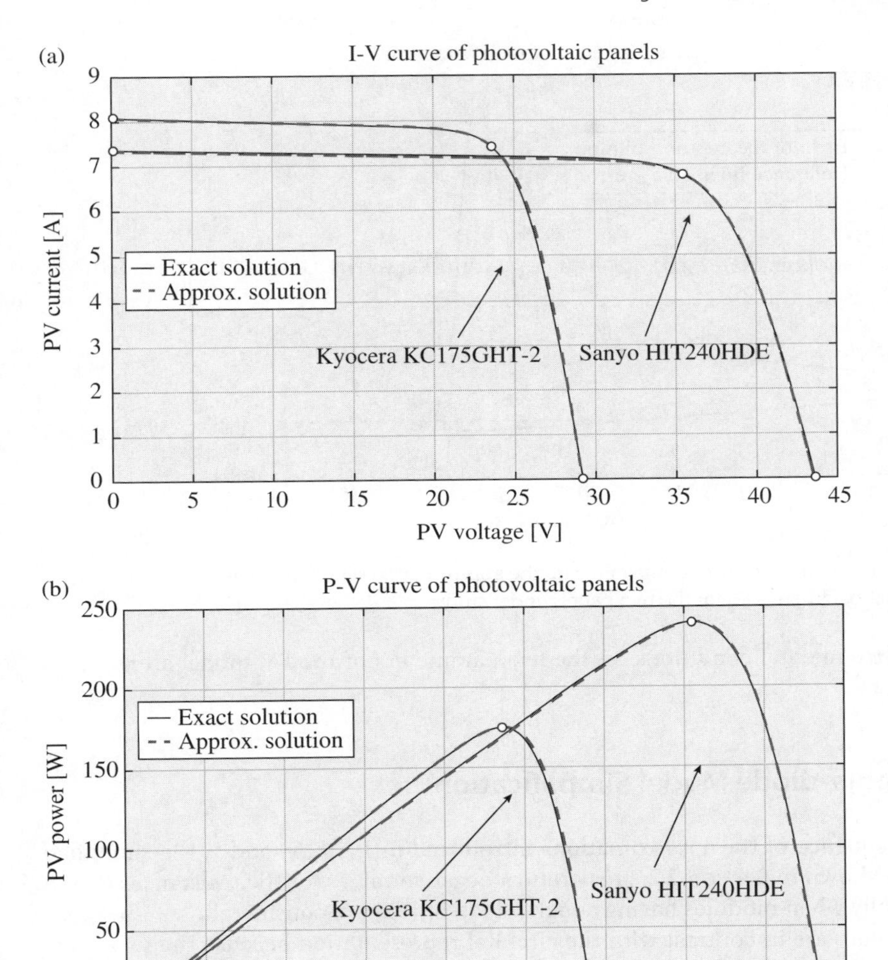

Figure 2.1 PV curves obtained using the approximated set of parameters (I_{ph}, I_s, η, R_s, R_h) and using an exact method [1] for the two panels under study.

rough procedure was used to import them, so part of the error might be ascribed to this procedure.

As shown in Table 2.4, the values of the five parameters change for different irradiance conditions. The analysis of such variations allows the effects of environmental changes occurring over the lifetime of the PV panel to be separated out, so that diagnostic analyses can be performed. Indeed, if the behavior of the five parameters as functions of the environmental conditions are known, any discrepancy of such parameters with respect to the reference values can be ascribed to an early degradation of the PV module's performance, so an alert can be activated. More details about the effect

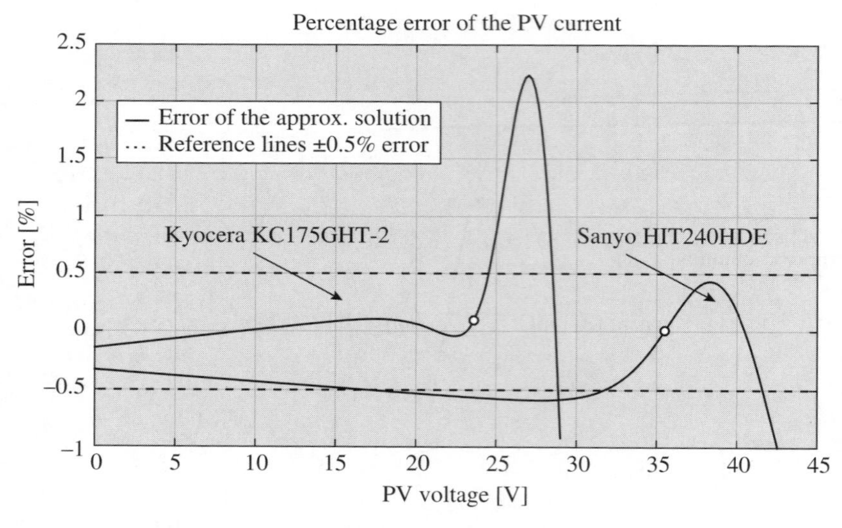

Figure 2.2 Error in the PV current obtained with the approximated set of parameters (I_{ph}, I_s, η, R_s, R_h) and the exact parameters [1] for the two panels under study.

of the environmental conditions on the five parameters of the PV model are given in Section 3.2.8.

2.3 Single-diode Model Simplification

As a consequence of the approximations introduced in I_{ph}, I_s, η, and x, the simplified procedure shown in Section 2.2.2 may return negative values for the resistances R_s and R_h, especially when modules having a high fill factor (FF) are considered.

These values are in contrast with the circuital representation because the two resistances are used to account for the losses inside the PV panels, thus assumed to be positive. Moreover, from a graphical point of view (see Section 1.4.1), it has been shown that the parameter R_s affects mainly the slopes of the I–V curve in the region between the MPP voltage and the open-circuit voltage, so values of R_s close to zero give I–V curves with highest slope at the right-hand side of the MPP. In contrast, R_h affects the slopes of the I–V curve in the region between the MPP current and the short-circuit current; I–V curves having an almost flat current region are characterized by R_h moving towards ∞. Thus by assuming that the I–V curve always has monotonical behavior, the two resistances can reach at most the limit values $R_s \rightarrow 0$ and $R_h \rightarrow \infty$.

A PV model having negative resistances does not have physical sense, this happening only because of the numerical approximations. For this reason, the negative values can be replaced with the corresponding limit values. From an analytical point of view this means that the corresponding resistances have no impact in the tuning of the SDM. This has been confirmed by several recent studies. For example, it has been demonstrated that, in many cases, an SDM characterized by only four parameters, obtained by removing one of the two resistances, is enough to reproduce accurately the I–V curves [17, 18]. Validation of such a simplified model has been carried out by means

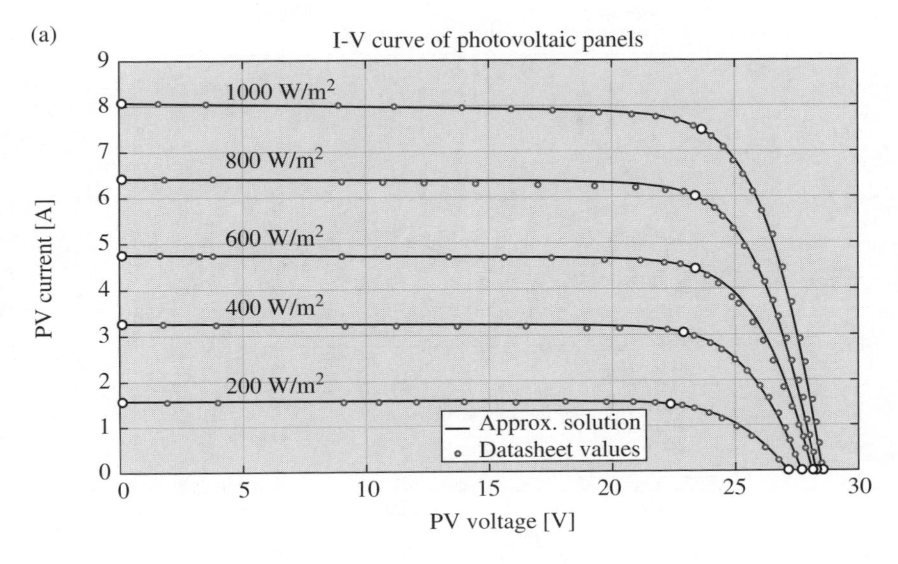

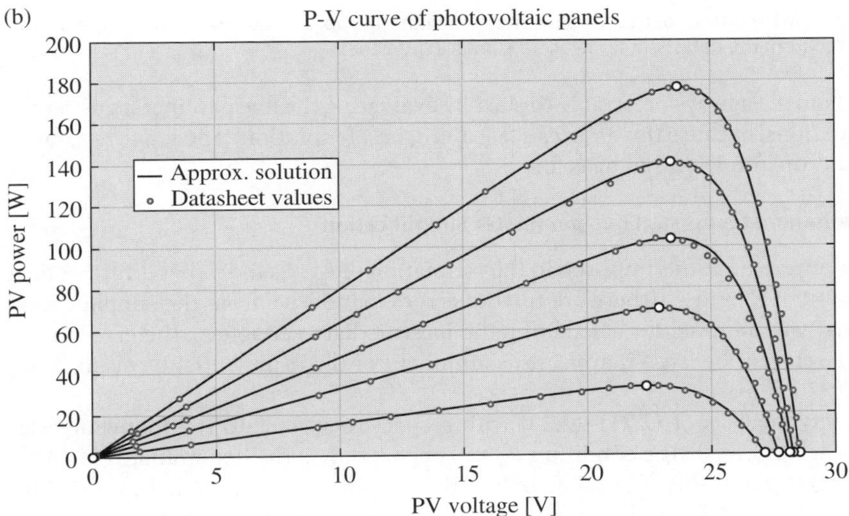

Figure 2.3 P–V and I–V curves for the Kyocera KC175GHT-2 PV panel; approximated parameters and experimental data extracted from the datasheet.

of an exhaustive analysis of many commercial panels [17]. Moreover, a sensitivity analysis has shown that, depending on the type of PV panel under study, only one resistance is dominant in the characterization of the SDM and that an indicator can be defined to determine, in a very simple way, the resistance that may be neglected [18, 19].

Unfortunately, in such papers the results of the analysis are shown only by referring to PV panels operating at STC, so it is not assured that the four-parameter model is suitable for reproducing the I–V curves in any environmental condition; a resistance that can be neglected at STC may be necessary to acheive a good fit at other operating conditions.

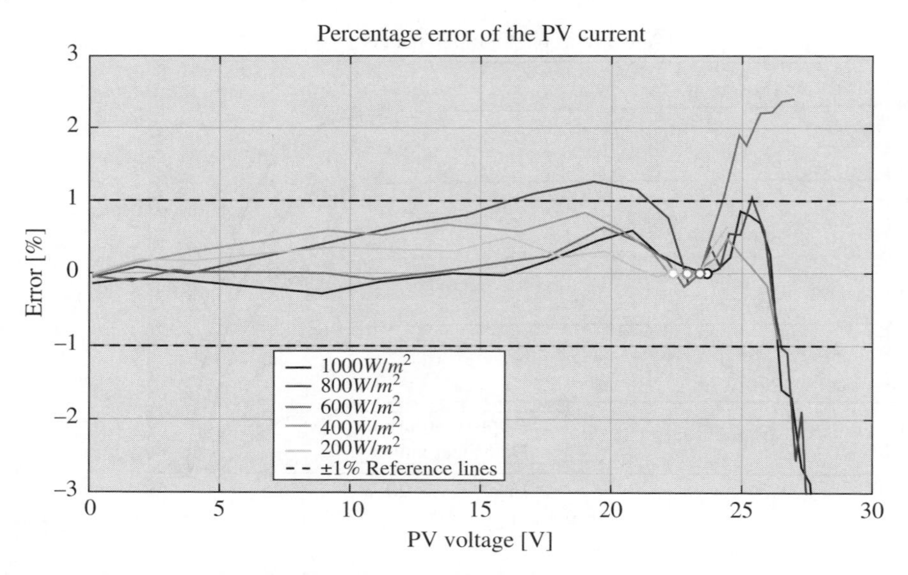

Figure 2.4 Error in PV current between value obtained with the approximated set of parameters and value extracted from the datasheet for Kyocera KC175GHT-2 panel.

For this reason it is always preferable to start by evaluating the five parameters using the explicit equations, because this assures a fast and accurate solution; if negative resistance values result, the model has to be refined.

2.3.1 Five-parameter versus Four-parameter Simplification

Besides the approximations imposed in the estimation of I_{ph}, I_s, and η that directly affect the evaluation of R_s and R_h, there are further errors that derive from the simplification in (2.27) that was adopted for calculating the intermediate variable x. The error in x is much more relevant because it appears as argument of the exponential terms in (2.23) and (2.24).

A sensitivity analysis of (2.21)–(2.23) with respect to x leads to the following relation: $R_s < 0$ when $x \rightarrow 0$. The condition $R_h < 0$ occurs only if the denominator of (2.23) becomes negative and this is possible only if the variable x assumes a high value. This correlation among R_s, R_h, and x ensures that, if x is not calculated correctly, the error

Table 2.4 Values of the five parameters of Kyocera KC175GHT-2 for different irradiance conditions.

	Irradiance [W/m²]				
	1000	**800**	**600**	**400**	**200**
I_s [nA]	2.072	2.205	2.388	2.673	3.241
I_{ph} [A]	8.083	6.421	4.76	3.271	1.575
R_s [Ω]	0.1331	0.1728	0.1884	0.2945	0.5968
R_h [Ω]	107.8	279.8	283.6	454.1	1463.9
η [/]	1.0499	1.0536	1.0662	1.0739	1.1012

leads to only one negative value for the resistances, so it is demonstrated that at most one resistance exceeds its physical limit, so it can be neglected.

Consequently, when the SDM is scaled down to four parameters, the procedure of Section 2.2.2 can be further simplified:

- Case A: $R_h < 0$ As $R_h \rightarrow \infty$, from (2.2) an explicit solution is given for x:

$$x = \ln\left(\frac{I_{ph} + I_s - I_{MPP}}{I_s}\right) \tag{2.33}$$

  Then recalculate R_s by using (2.21).
- Case B: $R_s < 0$
  Put $R_s = 0$. Then from (2.20) x is evaluated straightforwardly:

$$x = \frac{V_{MPP}}{\eta V_t} \tag{2.34}$$

  Then recalculate R_h by using (2.23).

Although the solutions obtained with these equations require that the real MPP matches the reconstructed curve, the condition $\partial P_{pv}/\partial V_{pv} = 0$ is not used for extracting the four parameters, so there is no assurance that the MPP calculated with the simplified model is very close to the real MPP. Other procedures [17, 18] might be also used for calculating the four parameters; some details about those methods are given in Section 2.3.2.

The occurrence of a negative resistance in the SDM will now be shown, using the Suntech STP280-24Vd PV panel as an example. For this commercial panel, it has been already shown that is better to use the simplified model with $R_h \rightarrow \infty$ [17]. Applying the procedure described in Section 2.2.2, as expected, the parallel resistance assumes a negative value, thus flagging that this model is not appropriate. Consequently R_h is neglected and (2.33) is used for updating R_s.

Table 2.5 shows the Suntech datasheet information used for evaluating the five-parameter and four-parameter solutions, which lead to the values reported in Table 2.6.

In Figure 2.5 the I–V and P–V curves obtained using the five- and four-parameter models are shown. The negative value of R_h in the five-parameter model produces a non-monotonical I–V curve (the continuous line in Figure 2.5a) and the maximum current does not correspond to the short-circuit current but it is located in the region just on the left-hand side of the MPP. Of course, this behavior is physically unacceptable. On the other side, the MPP (the square point in Figure 2.5), estimated using the

Table 2.5 Suntech STP280-24V datasheet information at STC.

	V_{oc} [V]	I_{sc} [A]	V_{MPP} [V]	I_{MPP} [A]	α_V [%/°C]	α_I [%/°C]
Suntech STP280-24Vd	44.8	8.33	35.2	7.95	−0.33	0.055

$G = 1000\ \mathrm{W/m^2}$ and $T = 25°\mathrm{C}$.

Table 2.6 Parameters of the SDM for the Suntech STP280-24Vd PV panel.

	Solution	
	5 parameters	4 parameters
I_s [nA]	$5.194 \cdot 10^{-1}$	$5.194 \cdot 10^{-1}$
I_{ph} [A]	8.243	8.243
R_s [Ω]	0.4681	0.3701
R_h [Ω]	**−270.13**	∞
η [/]	1.033	1.033

four-parameter model, does not perfectly match the experimental MPP (the white circle point in Figure 2.5). As expected, however, this difference is negligible and no further tuning of the values identified for the four parameters is required in order to achieve a better reconstruction of the I–V curve.

Figure 2.6 shows the percentage error of the current calculated as in (2.32). Comparing the currents obtained with the two models and the experimental data, it is evident that the four-parameter model fits much better.

2.3.2 Explicit Equations for Calculating the Four SDM Parameters

As mentioned in the previous section, in some cases, the five SDM parameters can be scaled down to four parameters without losing significant precision. The method introduced by Di Piazza et al. [18] allows PV modules to be classified into one of two groups on the basis of the value of a performance indicator called the serial–parallel ratio (*SPR*), defined as follows:

$$SPR = \frac{1 - \gamma_i}{e^{-r}} \tag{2.35}$$

where:

$$\gamma_i = \frac{I_{MPP}}{I_{sc}}; \quad \gamma_v = \frac{V_{MPP}}{V_{oc}}; \quad r = \frac{\gamma_i(1-\gamma_v)}{\gamma_v(1-\gamma_i)} \tag{2.36}$$

$SPR > 1$ indicates that the effect of R_s dominates, so the shunt resistance is neglected ($R_h \rightarrow \infty$). Conversely, $SPR < 1$ indicates that the effect of R_h is dominant and, as a consequence, the simplification $R_s = 0$ is valid.

The two non-negligible resistances can be calculated using the following equations:

$$R_s = \frac{V_{oc}}{I_{sc}} \frac{\frac{\gamma_v}{\gamma_i}(1-\gamma_i)ln(1-\gamma_i) + (1-\gamma_v)}{(1-\gamma_i)ln(1-\gamma_i) + \gamma_i} \quad \text{and} \quad R_h \rightarrow \infty \tag{2.37}$$

$$R_h = \frac{V_{oc}}{I_{sc}} \frac{\lambda_2 W\left(-SPR \cdot \lambda_1 e^{-\lambda_1}\right) + \lambda_1}{W\left(-SPR \cdot \lambda_1 e^{-\lambda_1}\right) + \lambda_1} \quad \text{and} \quad R_s = 0 \tag{2.38}$$

$W(\cdot)$ is the Lambert W-function, while λ_1 and λ_2 are given by:

$$\lambda_1 = \frac{1-\gamma_v}{1-\gamma_i} \cdot \frac{2\gamma_i - 1}{\gamma_i + \gamma_v - 1}; \quad \lambda_2 = \frac{\gamma_v}{1-\gamma_i} \tag{2.39}$$

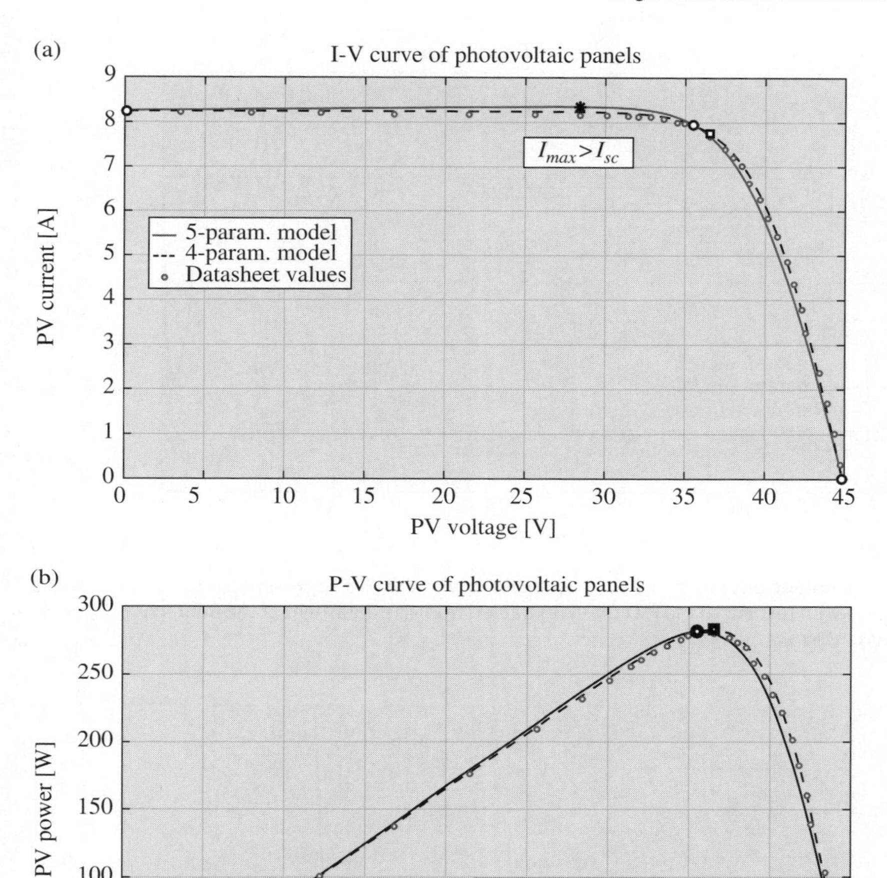

Figure 2.5 PV curves of the Suntech STP280-24Vd PV panel obtained using the two sets of approximated parameters.

With R_s and R_h known, the remaining parameters can be calculated as follows:

$$\eta = \frac{1}{N_s V_t} \frac{I_{MPP} R_s - (V_{oc} - V_{MPP})}{ln \frac{I_{sc} - I_{MPP} - \frac{V_{MPP}}{R_h}}{I_{sc} - \frac{V_{oc}}{R_h}}} \tag{2.40}$$

$$I_{ph} = I_{sc}; \quad I_s = \left(I_{ph} - \frac{V_{oc}}{R_h} \right) e^{-\frac{V_{oc}}{\eta N_s V_t}} \tag{2.41}$$

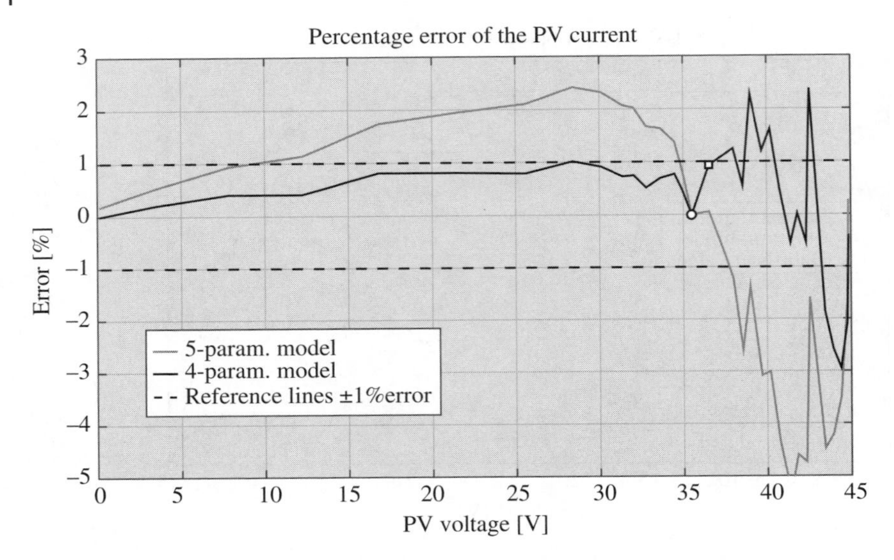

Figure 2.6 Percentage error in the PV current obtained with the two approximated sets of parameters (I_{ph}, I_s, η, R_s, R_h) with respect to the PV current extracted from the datasheet of the Suntech STP280-24Vd PV panel.

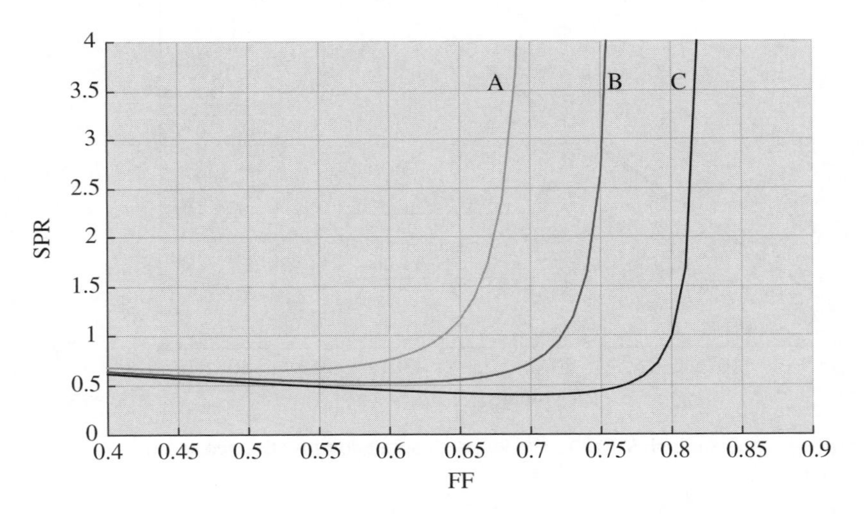

Figure 2.7 SPR indicator as a function of FF: (A) SPR_A at $\gamma_v = 0.75$; (B) SPR_B at $\gamma_v = 0.8$; (C) SPR_C at $\gamma_v = 0.85$.

It is worth noting that the equation used for calculating I_s does not depend on the energy gap of the semiconductor material, so it can also be applied to non-crystalline PV modules.

The SPR indicator can be also expressed as function of γ_v and the FF by using the identity $FF = \gamma_i \gamma_v$. The γ_v coefficient is usually constrained in the range [0.75, 0.85], so by fixing γ_v in that range the SPR can be plotted as function of FF only. Figure 2.7 shows that that PV panels having high FF will be categorized as $SPR > 1$, so for these PV panels the R_h parameter can be neglected. The complementary condition holds for a PV panel with low FF. The latter condition usually occurs for panels made with non-crystalline PV cells.

2.4 Improved Models for Amorphous and Organic PV Technologies

The accuracy of the SDM is adequate for crystalline and multi-crystalline silicon PV panels, but it may not be the right choice for other PV technologies. In this section, enhanced models used for PV panels based on amorphous and organic technologies will briefly be described. Amorphous PV panels are already gaining market share because they have features that are valuable, such as lightness and flexibility. Organic PV cells are still under investigation and many results have been obtained on laboratory prototypes. Nevertheless, the potential of the technology is very interesting and mass commercialization might be not so far off[2].

2.4.1 Modified SDM for Amorphous PV Cells

For amorphous PV technologies the additional intrinsic layer in the semiconductor region is the site of an intense recombination current that cannot be described with the SDM as it is. However, a current sink can be added to the traditional circuit, as shown in Figure 2.8 [13, 20]. Thus the SDM equation of (1.3) is modified as follows:

$$I_{pv} = I_{ph} - I_s \left(e^{\frac{V_{pv}+R_s \cdot I_{pv}}{\eta V t}} - 1 \right) - \frac{V_{pv} + R_s \cdot I_{pv}}{R_h} - \frac{\chi I_{ph}}{V_{bi} - (V_{pv} + R_s I_{pv})} \tag{2.42}$$

$$V_{bi} = N_s N_j V_c \tag{2.43}$$

The recombination current, described by the last term in (2.42), depends on the photo-induced current I_{ph} and the bias voltage $V_{pv} + R_s I_{pv}$. The V_{bi} term is the panel's built-in voltage, which is the product of the cell built-in voltage of a single junction V_c, the number of junctions in the cell N_j, and the number of cells in series N_s[3]. The χ coefficient depends on the intrinsic layer thickness d_i, which is usually an unknown parameter. The value of χ is therefore identified by means of a fitting procedure that maximizes the match between the I–V curve and the model, so it will be considered as a tunable coefficient used to improve the model accuracy[4].

As is evident, the new PV model includes the previous one for $\chi \to 0$. Although the new formulation is more general, it is not used for crystalline PV panels because it cannot be expressed in explicit form as the one in Section 1.5.1 can. Moreover, the presence of the voltage-controlled current generator in the modified SDM also requires the equations for calculating the $\overline{\mathbf{P}}$ parameters to be adapted. Of course, any numerical procedure based on curve-fitting and root-finding algorithms is applicable by forcing

---

2 The reader is invited to visit, for example, the web pages: http://depts.washington.edu/cmditr/modules/opv/organic_heterojunctions_in_solar_cells.html and http://plasticphotovoltaics.org/.

3 $V_c = 0.9$ V for amorphous silicon cells and, approximatively, $V_c = 0.6$ V for crystalline silicon cells. Usually N_j and N_s are provided by the PV manufacturers.

4 In Boyd et al. [13] the parameter $\chi = 6.07$ V is used for a two layers of amorphous silicon (2-a-Si) PV panels. $\chi = 0.0285$ V is used for single-layer mono-crystalline panel.

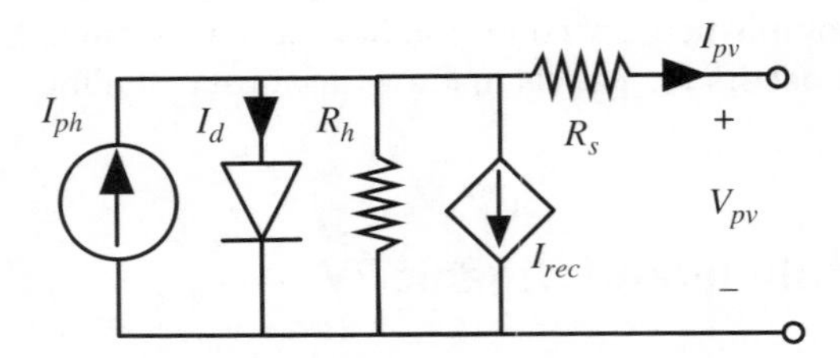

Figure 2.8 Single-diode equivalent circuit for amorphous PV panels.

the conditions (2.1)–(2.5), shown in Section 2.2.1, in Equation 2.8. Fortunately, in this case too, it is possible to calculate an approximate explicit solution that is a reasonable trade-off between simplicity and accuracy, as shown in the next section.

2.4.2 Five-parameter Calculation for Amorphous Silicon PV Panels

By evaluating (2.42) at the short-circuit condition, by assuming $R_h \to \infty$ and neglecting the diode current because it is in reverse operating condition, we get:

$$I_{ph} \simeq I_{sc} \frac{V_{bi}}{V_{bi} - \chi} \tag{2.44}$$

By exploiting (2.42) in the open-circuit condition, the parameter η is evaluated as:

$$\eta \simeq \frac{V_{oc}}{V_t \ln\left(\frac{I_ph - \chi I_{ph}/(V_{bi}-V_{oc})}{I_s}\right)} \tag{2.45}$$

The saturation current can again be calculated with (2.19), but now the γ_0 coefficient in (2.18) must be identified; by applying the same procedure of Femia et al. [7], the new coefficient is given by:

$$\gamma_0 = -\frac{V_{oc0}}{\alpha_v - \frac{V_{oc0}}{T_0}} \left(\frac{\alpha_I}{I_{sc0}} - \frac{3}{T_0} - \frac{E_{g0}}{kT_0^2} - \frac{\alpha_v \chi}{(V_{bi} - V_{oc})^2} \right) + \frac{E_{g0}}{kT_0} \tag{2.46}$$

The values R_s and R_h can be evaluated using the same intermediate variable x defined in (2.20). Consequently, R_s is still obtained straightforwardly by (2.21), while the new model formulation needs to evaluate the R_h resistance by evaluating (2.42) at the MPP:

$$R_h = \frac{x\eta V_t}{I_{ph} - I_{MPP} - I_s \cdot (e^x - 1) - \frac{\chi I_{ph}}{V_{bi} - x\eta V_t}} \tag{2.47}$$

The calculation of x is still performed by imposing the condition $\partial P_{pv}/\partial V_{pv} = 0$ at the MPP. The new expression for the current derivative with respect to the voltage, deduced from (2.42), is given by:

$$\left.\frac{\partial I_{pv}}{\partial V_{pv}}\right|_{MPP} = -\frac{\frac{1}{R_h} + \frac{I_s}{\eta \cdot V_t} \cdot e^{\frac{V_{MPP}+I_{MPP}\cdot R_s}{\eta V_t}} + \frac{\chi I_{ph}}{(V_{bi}-(V_{MPP}+R_s I_{MPP}))^2}}{1 + \frac{\chi I_{ph} R_s}{(V_{bi}-(V_{MPP}+R_s I_{MPP}))^2} + \frac{R_s}{R_h} + \frac{Rs \cdot I_s}{\eta \cdot V_t} \cdot e^{\frac{V_{MPP}+I_{MPP}\cdot R_s}{\eta V_t}}} \tag{2.48}$$

Substituting (2.20), (2.21), and (2.47) into (2.48) and using (2.5) leads to:

$$\frac{I_{MPP}}{V_{MPP}} + \frac{-\frac{e^x I_s}{\eta V_t} - \frac{\chi I_{ph}}{(V_{bi}-\eta V_t x)^2} - \frac{1}{\eta V_t x}\left(I^* - \frac{\chi I_{ph}}{V_{bi}-\eta V_t x}\right)}{1 + \frac{\eta V_t x - V_{MPP}}{\eta V_t I_{MPP}} \left(I_s e^x + \frac{\eta V_t \chi I_{ph}}{(V_{bi}-\eta V_t x)^2} + \frac{1}{x}\left(I^* - \frac{\chi I_{ph}}{V_{bi}-\eta V_t x}\right)\right)} = 0 \tag{2.49}$$

where $I^* = I_{ph} - I_{MPP} - I_s(e^x - 1)$.

Also in this case, the analysis of the terms appearing in (2.25) is used to simplify the equations. In particular, the term $(V_{bi} - \eta V_t x)^2$ can be simplified as follows:

$$\begin{aligned}(V_{bi} - \eta V_t x)^2 &\simeq V_{bi}^2 + \eta V_t \left(2V_{MPP}x - \frac{V_{MPP}^2}{\eta V_t}\right) - 2\eta V_t x V_{bi} \\ &= V_{bi}^2 + V_{MPP}^2 - 2V_{bi}V_{MPP} - 2R_s I_{MPP}(V_{MPP} - V_{bi}) \\ &\simeq V_{bi}^2 + V_{MPP}^2 - 2V_{bi}V_{MPP} = (V_{bi} - V_{MPP})^2 = \beta \end{aligned} \tag{2.50}$$

The quantity $2R_s I_{MPP}(V_{MPP} - V_{bi})$ can be neglected because it is significantly smaller than the other terms.

After some manipulations and simplifications in (2.49), the following equation is obtained:

$$e^x I_s \left(2 - \frac{V_{MPP}}{\eta V_t}\right) = I_{ph} + I_s - 2I_{MPP} + \chi I_{ph}\left(\frac{V_{MPP}}{\beta} - \frac{1}{V_{bi} - V_{MPP}}\right) \tag{2.51}$$

It is worth noting that to obtain (2.51), the approximation $\eta V_t x = V_{MPP} + R_s I_{MPP} \simeq V_{MPP}$ has also been used in (2.49). This simplification is valid because $R_s I_{MPP}$ is usually significantly lower than V_{MPP}. Finally, the variable x is calculated as:

$$x = \ln\left(\frac{I_{ph} + I_s - 2I_{MPP} + \chi I_{ph}\left(\frac{V_{MPP}}{\beta} - \frac{1}{V_{bi} - V_{MPP}}\right)}{I_s\left(2 - \frac{V_{MPP}}{\eta V_t}\right)}\right) \tag{2.52}$$

The equations have been validated for the Solarex MST-56MV PV module. This module is fabricated using an advanced tandem-junction thin-film technology. The tandem-junction structure is obtained by stacking two solar cells vertically. The two cells are different and are tuned for the optimal conversion of different segments of the irradiance spectrum. Table 2.7 shows the Solarex datasheet information used for evaluating the five parameters of the SDM. Table 2.8 reports the values of the five parameters used in the SDM equation (1.3) and the ones used in (2.42) for accounting for the recombination current. For both cases, the sets of explicit equations have been used to calculate I_{ph}, I_s, η, R_s, and R_h.

The data in Table 2.8 are used to generate the curves in Figure 2.5. The results obtained with the model accounting for the recombination current are very close to the experimental data extracted from the datasheet, demonstrating the effectiveness of the improved model. At the same time this good agreement of the improved

Table 2.7 Datasheet information for Solarex MST-56MV at STC.

	V_{oc} [V]	I_{sc} [A]	V_{MPP} [V]	I_{MPP} [A]	α_V [mV/°C]	α_I [mA/°C]
Solarex MST-56MV	102	0.871	73	0.761	−0.492	0.0088

$G = 1000\ \mathrm{W/m^2}$ and $T = 25°\mathrm{C}$. Solarex MST-56MV has 71 serially connected tandem-junction cells, thus $N_s = 72$ and $N_j = 2$.

Table 2.8 Calculated SDM parameters for the Solarex MST-56MV.

	Standard model parameters	Improved model parameters
I_s [nA]	0.598	0.753
I_{ph} [A]	0.859	0.901
R_s [Ω]	20.648	21.482
R_h [Ω]	789.2	3767
η [/]	2.607	2.661

model confirms the validity of the simplified procedure for calculating the five parameters that has been developed in this section. Further details concerning the procedure for calculating the parameters of amorphous PV modules are reported in the literature [21].

2.4.3 Modified Model for Organic PV Cells

Organic solar cells are a promising alternative to conventional inorganic PV devices because they offer many advantages: low-production cost due to a simpler fabrication process, integration on unconventional substrates, and the existence of organic structures whose functionalities can be chemically tuned to adjust the energy levels and therefore improve light absorption and charge transport. The most thoroughly investigated kind of organic solar cells use a bulk heterojunction (BHJ), obtained as a mixture of two materials, one an electron donor (polymer) and the other an electron acceptor (such as fullerene). The role of the BHJ is to create a donor/acceptor (D/A) interface that allows the transfer of the photo-induced charge. In contrast to silicon solar cells, in organic PVs the electron–hole pair created by an incoming photon is strongly bound in a so-called "exciton". To enable utilization of the charge carriers the exciton has to be dissociated and this happens at the D/A interface.

BHJ solar cells have a characteristic S-shape in the I–V curve. The S-shape has a detrimental effect on the FF and consequently on cell efficiency. The origin of the I–V curve deformation is currently under investigation and many efforts have been made to reduce its detrimental effect. Power conversion efficiencies of up to 10% have already been reached for a tandem cell geometry [22].

From the electrical point of view, the S-shape cannot be reproduced by using the SDM as it is, so an additional non-linear block is introduced, as shown in Figure 2.10. For organic PV technology, the electrical model is used for reproducing the I–V curve of a single cell because, to the best of the authors' knowledge, organic PV modules consisting of single cells are not yet available in the market. It is worth noting that, for laboratory I–V characterization, it is common practice to consider the PV cell as a load, so its active region is in the fourth quadrant of the I–V plane, if the current and voltage references given in Figure 2.10 are considered. Referring to the scheme of Figure 2.10, the total voltage at the PV terminals is:

$$V_{pv} = R_s I_{pv} + V_1 + V_2 \tag{2.53}$$

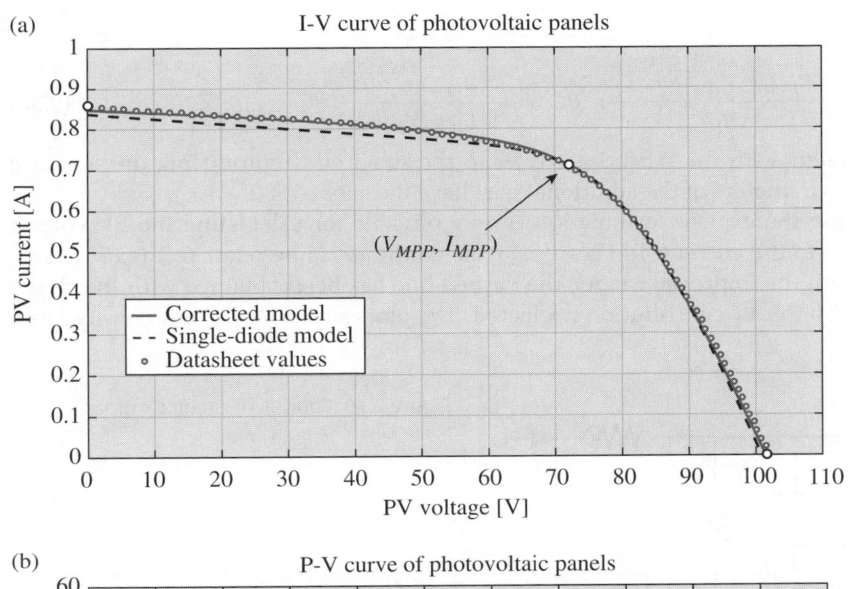

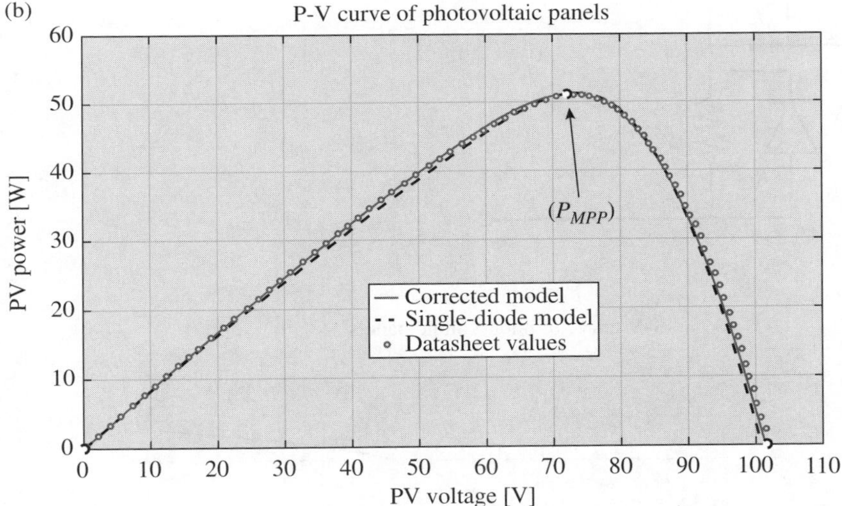

Figure 2.9 PV curves of the Solarex MST-56MV PV panel obtained by using the standard SDM and the improved model [13].

According to the results shown in Romero et al. [23], the voltages V_1 and V_2 of the two non-linear blocks can be expressed explicitly with respect to the current by using the Lambert W-function:

$$V_1 = \eta_1 V_t \cdot \ln \left[\frac{\eta_1 V_t}{R_{h1} I_{s1}} \cdot W\left(\theta_{V1}\right) \right] \tag{2.54}$$

$$V_2 = -\eta_2 V_t \cdot \ln \left[\frac{\eta_2 V_t}{R_{h2} I_{s2}} \cdot W\left(\theta_{V2}\right) \right] \tag{2.55}$$

where:

$$\theta_{V1} = \frac{R_{h1} I_{s1} \cdot e^{\frac{R_{h1}(I_{ph}+I_{s1}+I_{pv})}{\eta_1 V_t}}}{\eta_1 V_t} \qquad \theta_{V2} = \frac{R_{h2} I_{s2} \cdot e^{\frac{R_{h2}(I_{pv}-I_{s1})}{\eta_2 V_t}}}{\eta_2 V_t} \tag{2.56}$$

The parameters with the subscript 1 refer to the subcircuit reproducing the standard model; the parameters of the additional part have the subscript 2.

In this case the explicit formulation is only possible for calculating the PV voltage with respect to the current. In Figure 2.11 the continuous line refers to the PV curves obtained with the improved model; the dashed line has been obtained with the classical SDM with the V_2 contribution neglected. The plots show the current density J as a

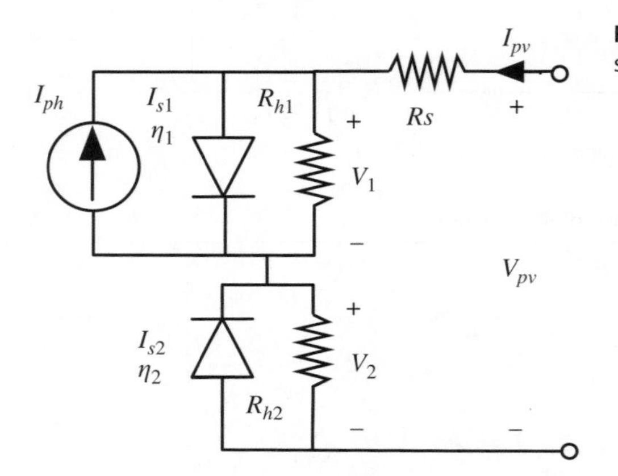

Figure 2.10 Equivalent circuit for organic solarcells.

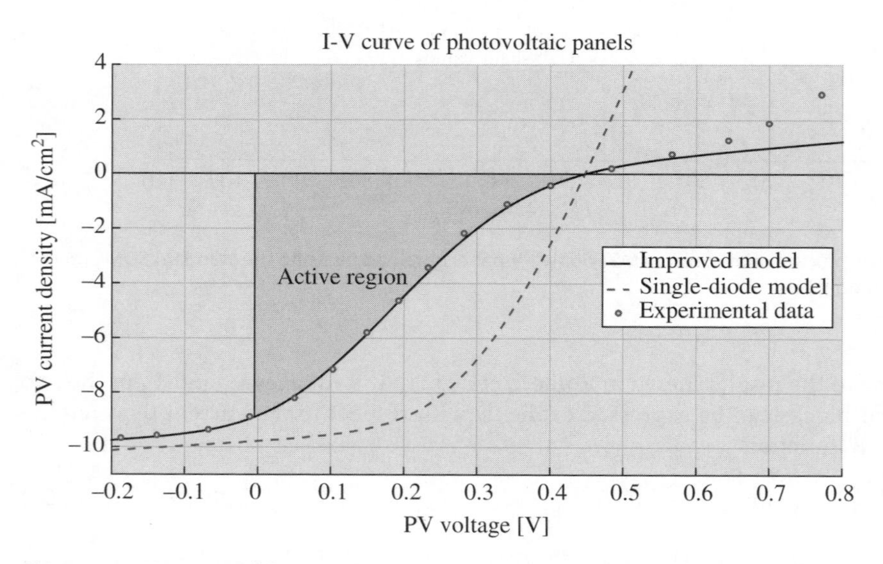

Figure 2.11 PV curves of the organic PV cell obtained using standard SDM and the improved version [23].

Table 2.9 Parameters of the two-diode model for an organic PV cell.

J_{ph} $\left[\frac{A}{m^2}\right]$	R_s $[\Omega m^2]$	R_{h1} $[\Omega m^2]$	R_{h2} $[\Omega m^2]$	η_1 $[/]$	η_2 $[/]$	I_{s1} $[A/m^2]$	I_{s2} $[A/m^2]$
100	1.25×10^{-3}	5.8×10^{-2}	4.1×10^{-2}	1.9	2.55	9.5×10^{-3}	3.85

function of the PV cell voltage; the numerical results are compared with experimental data [24]. The parameters used are shown in Table 2.9.

References

1 Brano, V.L. and Ciulla, G. (2013) An efficient analytical approach for obtaining a five parameters model of photovoltaic modules using only reference data. *Applied Energy*, **111** (0), 894–903, doi:10.1016/j.apenergy.2013.06.046. URL: http://www.sciencedirect.com/science/article/pii/S0306261913005539.

2 Ishaque, K. and Salam, Z. (2011) An improved modeling method to determine the model parameters of photovoltaic (PV) modules using differential evolution (DE). *Solar Energy*, **85** (9), 2349–2359, doi:10.1016/j.solener.2011.06.025. URL: http://www.sciencedirect.com/science/article/pii/S0038092X11002374.

3 Karatepe, E., Boztepe, M., and Colak, M. (2006) Neural network based solar cell model. *Energy Conversion and Management*, **47** (9–10), 1159–1178, doi:10.1016/j.enconman.2005.07.007. URL: http://www.sciencedirect.com/science/article/pii/S019689040500186X.

4 Tian, H., Mancilla-David, F., Ellis, K., Muljadi, E., and Jenkins, P. (2012) A cell-to-module-to-array detailed model for photovoltaic panels. *Solar Energy*, **86** (9), 2695–2706, doi:10.1016/j.solener.2012.06.004. URL: http://www.sciencedirect.com/science/article/pii/S0038092X12002150.

5 Chan, D. and Phang, J. (1987) Analytical methods for the extraction of solar-cell single- and double-diode model parameters from I–V characteristics. *Electron Devices, IEEE Transactions on*, **34** (2), 286–293, doi:10.1109/T-ED.1987.22920.

6 Eicker, U. (2003) *Solar Technologies for Buildings*, John Wiley.

7 Femia, N., Petrone, G., Spagnuolo, G., and Vitelli, M. (2013) *Power Electronics and Control Techniques for Maximum Energy Harvesting in Photovoltaic Systems*, 1st edn. CRC Press.

8 Sera, D., Teodorescu, R., and Rodriguez, P. (2007) PV panel model based on datasheet values, in *2007 IEEE International Symposium on Industrial Electronics*, pp. 2392–2396.

9 Orioli, A. and Di Gangi, A. (2013) A procedure to calculate the five-parameter model of crystalline silicon photovoltaic modules on the basis of the tabular performance data. *Applied Energy*, **102** (0), 1160–1177.

10 Ramos-Paja, C.A., Petrone, G., and Spagnuolo, G. (2013) Symbolic algebra for the calculation of the series and parallel resistances in PV module model, in *International Conference on CLEAN ELECTRICAL POWER Renewable Energy Resources Impact (ICCEP)*, Alghero, Sardinia, Italy, pp. 62–66.

11 De Soto, W., Klein, S., and Beckman, W. (2006) Improvement and validation of a model for photovoltaic array performance. *Solar Energy*, **80** (1), 78–88.
12 Liu, S. and Dougal, R. (2002) Dynamic multiphysics model for solar array. *Energy Conversion, IEEE Transactions on*, **17** (2), 285–294, doi:10.1109/TEC.2002.1009482.
13 Boyd, M.T., Klein, S.A., Reindl, D.T., and Dougherty, B.P. (2011) Evaluation and validation of equivalent circuit photovoltaic solar cell performance models. *Journal of Solar Energy Engineering*, **133** (2), 021 005.
14 Californian Energy Commission. Guidelines for California's solar electric incentive programs, 2nd edn. CEC-300-2008-007-CMF, State of California Energy Commission.
15 Weisstein, E. (2008) Lambert W-function. *A Wolfram Web Resource*. http://mathworld.wolfram.com/LambertW-Function.html.
16 Corless, R., Gonnet, G., Hare, D., Jeffrey, D., and Knuth, D. (1996) On the Lambert W function. *Advances in Computational Mathematics*, **5** (1), 329–359, doi:10.1007/BF02124750.
17 Mahmoud, Y., Xiao, W., and Zeineldin, H. (2013) A parameterization approach for enhancing PV model accuracy. *Industrial Electronics, IEEE Transactions on*, **60** (12), 5708–5716.
18 Cannizzaro, S., Di Piazza, M., Luna, M., and Vitale, G. (2014) Generalized classification of PV modules by simplified single-diode models, in *Industrial Electronics (ISIE), 2014 IEEE 23rd International Symposium on*, pp. 2266–2273, doi:10.1109/ISIE.2014.6864971.
19 Di Piazza, M., Luna, M., Petrone, G., and Spagnuolo, G. (2015) About the identification of the single-diode model parameters of high-fill-factor photovoltaic modules, in *Clean Electrical Power (ICCEP), 2015 International Conference on*, pp. 85–91, doi:10.1109/ICCEP.2015.7177605.
20 Merten, J., Asensi, J.M., Voz, C., Shah, A., Platz, R., and Andreu, J. (1998) Improved equivalent circuit and analytical model for amorphous silicon solar cells and modules. *Electron Devices, IEEE Transactions on*, **45** (2), 423–429, doi:10.1109/16.658676.
21 Petrone, G. and Spagnuolo, G. (2015) Parameters identification of the single-diode model for amorphous photovoltaic panels, in *Clean Electrical Power (ICCEP), 2015 International Conference on*, pp. 105–109, doi:10.1109/ICCEP.2015.7177608.
22 Zuo, L., Yao, J., Li, H., and Chen, H. (2014) Assessing the origin of the s-shaped I–V curve in organic solar cells: an improved equivalent circuit model. *Solar Energy Materials and Solar Cells*, **122** (0), 88–93, doi:10.1016/j.solmat.2013.11.018. URL: http://www.sciencedirect.com/science/article/pii/S0927024813006028.
23 Romero, B., del Pozo, G., and Arredondo, B. (2012) Exact analytical solution of a two diode circuit model for organic solar cells showing s-shape using Lambert W-functions. *Solar Energy*, **86** (10), 3026–3029, doi:10.1016/j.solener.2012.07.010. URL: http://www.sciencedirect.com/science/article/pii/S0038092X12002526.
24 García-Sánchez, F., Lugo-Munoz, D., Muci, J., and Ortiz-Conde, A. (2013) Lumped parameter modeling of organic solar cells' s-shaped I–V characteristics. *Photovoltaics, IEEE Journal of*, **3** (1), 330–335, doi:10.1109/JPHOTOV.2012.2219503.

3

PV Simulation under Homogeneous Conditions

3.1 Introduction

The five-parameter model is of interest because it requires only a small amount of input data, usually provided by the manufacturer, and therefore it is a valuable tool for simulating the PV field. As shown in the previous chapter, there are many approaches to identifying the values of the set of parameters (I_{ph}, I_s, η, R_s, and R_h) that guarantee that the equivalent electrical circuit matches the I–V curve. The parameters can be easily calculated if the short-circuit current, the open-circuit voltage and the voltage and current at the MPP are known, so the SDM is able to accurately represent the PV electrical behavior for any environmental conditions provided that different sets of the five-parameters are used for different values of G and T.

Unfortunately the PV module datasheet typically provides the values of the electrical parameters only at STC, so that the SDM allows to predict the I–V curve in any environmental condition. The dependency of the SDM parameters on environmental conditions is another aspect which is widely treated in the literature, from which it is evident that errors in parameter estimation are transferred to invalid predictions of PV performance. Significant efforts have been made to obtain additional equations that define the way in which the SDM parameters change with respect to the irradiance and temperature of the PV cells. Some authors suggest keeping some of the five parameters at a constant value [1], and to change others that are related, through "translating" formulas to varying environmental conditions [2, 3].

In this chapter, the most popular approaches will be described. The translating equations, in combination with the systematic procedure used for extracting the five (or four) parameters, as described in Chapter 2, allow the SDM model to be tuned to reproduce the behavior of PV panels in any environmental conditions.

3.2 Irradiance- and Temperature-dependence of the PV Model

3.2.1 Direct Effects of Irradiance and Temperature

Irradiance and ambient temperature variations affect the behavior of the PV module in a very complex way; some direct effects can be deduced from physical relations, but others can be only approximated using empirical equations.

Photovoltaic Sources Modeling, First Edition. Giovanni Petrone, Carlos Andrés Ramos-Paja and Giovanni Spagnuolo.

Companion Website: www.wiley.com/go/petrone/Photovoltaic_Sources_Modeling

Some dependencies have been already described; an example is the equation describing the saturation current I_s, which is strongly dependent on the cell temperature, as shown in (2.16). The effect of the irradiance on I_s, in turn, has been described using an empirical correction coefficient, as defined in (2.19).

The photo-induced current, representing the quantity of electron–hole pairs responsible for the current flowing in the PV cell, is directly proportional to the number of photons arriving at the PV surface having an energy higher than the energy bandgap E_g, so the expression for I_{ph} can be easily related to the irradiance value by using the following expression [4]:

$$I_{ph}(G,T) = \overline{R} \cdot A \cdot G + \delta I_{ph}(T) \tag{3.1}$$

where A is the active area, G is the solar irradiance, and $\overline{R}$ is the spectral-averaged responsivity of the PV cells (A/W). $\delta I_{ph}(T)$ is a thermal coefficient, accounting for the temperature-dependence of the photo-induced current. The spectral-averaged responsivity is given by:

$$\overline{R} = \frac{\int_0^{\lambda(E_g)} R(\lambda)E(\lambda)d\lambda}{\int_0^{\infty} E(\lambda)d\lambda} \tag{3.2}$$

where $E(\lambda)$ is the solar irradiance spectral distribution function and $\lambda(E_g)$ is the wavelength corresponding to the energy bandgap E_g. The responsivity $R(\lambda)$ depends not only on the cell material, but also on the weather conditions since the wavelength distribution of the irradiance varies as sunlight passes through the atmosphere.

The variation of the cell temperature causes two contrasting effects on I_{ph}. The first one is that the energy bandgap decreases at higher temperatures, so the number of photons that can produce electron–hole pairs increases; on the other side higher temperatures reduce the diffusion length and lifetime of carriers, so the recombination losses increase. The net effect is that the photo-induced current slightly increases. Its variation is approximated with a linear function, so (3.1) becomes:

$$I_{ph}(G,T) = \overline{R} \cdot A \cdot G + \alpha_i (T - T_0)\frac{G}{G_0} \tag{3.3}$$

The subscript "0" indicates that the corresponding parameter has been calculated for the reference (for example, STC) condition. In order to avoid calculating the physical parameters appearing in (3.3), the last term might be posed in a simpler form by exploiting the value of the photo-induced current $I_{ph,0}$ for a known irradiance condition G_0:

$$I_{ph} = \frac{G}{G_0}\left[I_{ph,0} + \alpha_i (T - T_0)\right] \tag{3.4}$$

The irradiance G represents the effective irradiance on the PV cell surface, by assuming that it is already corrected for any losses caused by environmental and operating conditions (for example reflection losses due to the incident angle of the solar radiation, dirt on the PV surface, and so on)[1]. Assuming the condition $I_{ph} \simeq I_{sc}$, Equation 3.4 is also valid for the short-circuit current; the temperature coefficient α_i is provided by the manufacturer in the PV panel datasheet.

1 The reader is invited to refer to Section 3.3.1 for further details of the factors affecting the effective irradiance G and how to calculate it.

Many papers have shown that the relationship between the open-circuit voltage V_{oc} and the irradiance is logarithmic, but the effect of the temperature on V_{oc} is almost linear, due to the exponential increase of the saturation current with temperature. The explicit relationship can be deduced by assuming the ideal equation of the PV model, obtained by neglecting R_s and R_h in (1.3) [5]. This approximation leads to the following expression for the open-circuit voltage:

$$V_{oc} = \eta V_t \ln\left(1 + \frac{I_{ph}}{I_s}\right) \simeq \eta V_t \ln\left(\frac{I_{ph}}{I_s}\right) \tag{3.5}$$

At a given irradiance value, the temperature coefficient of the open-circuit voltage can be expressed as:

$$\left.\left(\frac{\partial V_{oc}}{\partial T}\right)\right|_{G=G_0} = \frac{V_{oc}(G_0, T) - V_{oc}(G_0, T_0)}{T - T_0} \simeq \frac{\eta V_t \ln\left(\frac{I_{ph}(G_0,T)}{I_s(G_0,T)}\right) - V_{oc0}}{T - T_0} \tag{3.6}$$

where $V_{oc}(G_0, T)$ has been replaced with (3.5) and the open-circuit voltage in (G_0, T_0) corresponds to V_{oc0}. From (3.6), by adding on both sides the value of the open-circuit voltage $V_{oc}(G, T)$ under arbitrary irradiance and temperature conditions, the following expression holds:

$$V_{oc}(G, T) + (T - T_0)\left.\left(\frac{\partial V_{oc}}{\partial T}\right)\right|_{G_0} \simeq V_{oc}(G, T) + \eta V_t \ln\left(\frac{I_{ph}(G_0, T)}{I_s(G_0, T)}\right) - V_{oc0} \tag{3.7}$$

Using (3.5) to express $V_{oc}(G, T)$ and by performing some rearrangements, the previous equation becomes:

$$V_{oc}(G, T) \simeq (T - T_0)\left.\left(\frac{\partial V_{oc}}{\partial T}\right)\right|_{G_0} + \eta_0 V_t \ln\left(\frac{\frac{I_{ph}(G,T)}{I_s(G,T)}}{\frac{I_{ph}(G_0,T)}{I_s(G_0,T)}}\right) + V_{oc0} \tag{3.8}$$

where the condition $\eta = \eta_0$ is assumed to hold[2]. Finally, by using the normalized temperature coefficient $\alpha_v = \frac{1}{V_{oc0}} \frac{\partial V_{oc}}{\partial T}$ calculated at STC, the previous equation is simplified as:

$$V_{oc}(G, T) \simeq V_{oc0}\left[1 + \alpha_v(T - T_0)\right] + \eta_0 V_t \ln\left(\frac{\frac{I_{ph}(G,T)}{I_s(G,T)}}{\frac{I_{ph}(G_0,T)}{I_s(G_0,T)}}\right) \tag{3.9}$$

Equations 2.16 and 2.19 show that the saturation current is almost independent of the irradiance: $I_s(G, T) \simeq I_s(G_0, T)$, so the substitution of (3.4) into (3.9) leads to:

$$V_{oc}(G, T) \simeq V_{oc0}\left[1 + \alpha_v(T - T_0) + \frac{\eta_0 V_t}{V_{oc0}} \ln\left(\frac{G}{G_0}\right)\right] \tag{3.10}$$

Although (3.10) has been obtained by performing more than one approximation, it has been widely investigated and experimentally validated [6, 7]. Such documents highlight

2 The scientific community has contrasting opinions about the assumption that η does not change with the environmental conditions. But it is reasonable to suppose a limited variation for η, so its variation can be neglected in the estimation of V_{oc} and the value calculated at STC (η_0) can be used.

that (3.10) and (3.4) are reliable for translating the corresponding parameters to any environmental condition, provided that the temperature coefficients and the cell temperature are evaluated correctly.

In outdoor conditions the cell temperature T cannot be measured directly because the cell surface is not accessible, so a correct estimation is mandatory for ensuring correct calculation of the PV parameters. The theoretical calculation of T requires the use of thermal models based on the reflectivity and transmittivity properties of the PV panel, and these parameters are not always reported in the datasheet and are not easy to measure, thus empirical and semi-empirical methods are used for their simplicity and because they use parameters that are available through the manufacturer. Examples of detailed PV thermal models can be found in the literature [8, 9], as well as various semi-empirical procedures [10, 11].

The empirical solution proposed by King et al. [12] is a widely used approach to estimate the cell temperature T (see Equation 3.11). The temperature at the back of the PV modules T_m and the global irradiance G_T must be measured or estimated in order to calculate T. If not available, T_m is obtained from (3.12), which involves the ambient temperature T_a and the wind speed (WS) measured at standard 10 m height. The coefficients a, b, and ΔT are reported in Table 3.1 for different operating conditions.

$$T = T_m + \frac{G_T}{G_{ref}} \cdot \Delta T \tag{3.11}$$

$$T_m = T_a + \frac{G_T}{G_{ref}} \cdot \left[1000 \cdot e^{(a+b \cdot WS)}\right] \tag{3.12}$$

As an alternative, an extremely simple equation used for estimating the cell temperature is the one involving the "nominal operating cell temperature" (NOCT) coefficient[3]. The latter is always available in the panel datasheet and relates the cell temperature to the environmental conditions as follows:

$$T = T_a + \frac{NOCT - 20}{800} G \tag{3.13}$$

Table 3.1 Empirical coefficients for estimating cell temperature T and temperature at the rear of the PV panel.

Module type	**Mounting conditions**	***a***	***b* [s/m]**	**ΔT [°C]**
Glass/cell/glass	Open rack	−3.47	−0.0594	3
Glass/cell/glass	Close roof mount	−2.98	−0.0471	1
Glass/cell/polymer sheet	Open rack	−3.56	−0.0750	3
Glass/cell/polymer sheet	Close roof mount	−2.81	−0.0455	0
Polymer/thin-film/stell	Open rack	−3.58	−0.113	3
22X linear concentrator	Tracker	−3.23	−0.13	13

Source: King et al. [12].

3 The NOCT coefficient is measured in the operating conditions fixed at irradiance of 800 W/m^2, ambient temperature of 20°C, and wind speed of 1 m/s.

Of course, Equation 3.13 does not consider the mounting condition and the wind speed, which are two parameters influencing the cell temperature in outdoor operation. An exhaustive survey of equations developed for estimating the operating temperature of PV panels is found in the paper by Skoplaki and Palyvos [13].

The dependence of the remaining parameters of the SDM model with respect to environmental conditions is not yet well established and new equations are continuously being proposed in the literature. In the following discussion the most widely accepted approaches will be described and compared, so as to show the main advantages and drawbacks of each.

3.2.2 Equations for "Translating" SDM Parameters

In addition to Equation 3.4, the model proposed by De Soto et al. uses the following equations to express the variations of the SDM parameters as a function of cell temperature and irradiance [1]:

$$I_s = I_{s0}\left(\frac{T}{T_0}\right)^3 e^{\left[\frac{1}{k}\left(\left.\frac{E_g}{T}\right|_{T_0} - \frac{E_g}{T}\right)\right]} \tag{3.14}$$

$$\frac{E_g(T)}{E_g(T_0)} = 1 - 0.0002677(T - T_0) \tag{3.15}$$

$$R_s = R_{s0} \tag{3.16}$$

$$R_h = \frac{G_0}{G} R_{h0} \tag{3.17}$$

$$\eta = \eta_0 \tag{3.18}$$

The equation describing the saturation current is the same as already introduced in (2.16), but it is expressed as function of the reference value I_{s0} instead of the C_0 coefficient.

The temperature dependence of the material energy bandgap E_g is accounted for by means of the linear equation (3.15). The ideality factor η and the series resistance R_s have been considered constant. The shunt resistance R_h is assumed to be inversely proportional to the irradiance G (3.17).

The primary advantage of the De Soto model is that it requires the five parameters to be calculated only in a reference condition, so the information given in the PV module datasheet can be processed by any of the procedures described previously for calculating the parameters I_{ph0}, I_{s0}, R_{s0}, R_{sh0}, and η_0, and then the "translation" equations of (3.4)–(3.18) give the parameter values in any other environmental condition.

The procedure has been applied for estimating the five parameters of the Kyocera KC175GHT-2 at medium ($G = 400\ \text{W/m}^2$) and low ($G = 200\ \text{W/m}^2$) irradiance conditions. The cell temperature has been considered constant ($T = 25°\text{C}$) in this example.

In Table 3.2, the columns headed "Translated solution" give the values of I_{ph}, I_s, η, R_s, and R_h obtained by applying the previous equations, starting from the parameter values calculated at STC and already shown in Table 2.3. The columns headed "Approx. solution" give the five parameters calculated by using the procedure shown in Section 2.2.2. This approach uses the datasheet values for the short-circuit current, the open-circuit voltage and the MPP voltage and current in the irradiance conditions under test to calculate the SDM parameters.

Table 3.2 Values of the five parameters of Kyocera KC175GHT-2.

	$G = 400\,W/m^2$		$G = 200\,W/m^2$	
	Translated solution	**Approx. solution**	**Translated solution**	**Approx. solution**
I_s [nA]	1.0660×10^{-1}	2.673	1.0660×10^{-1}	3.241
I_{ph} [A]	3.244	3.271	1.622	1.575
R_s [Ω]	0.2836	0.2945	0.2836	0.5968
R_h [Ω]	208.25	454	416.5	1463
η [/]	0.9474	1.0739	0.9474	1.1012

In Figure 3.1 the I–V curves obtained with the two sets of parameters have been compared with the experimental results taken from the datasheet. The I–V curve, calculated by using the De Soto parameters (the continuous lines in the figure), has a steep slope on the right of the MPP and a strong discrepancy with respect to the experimental data, especially in the estimation of the MPP. It is well known that the R_s parameter has an effect on that slope, but in De Soto's translating equations R_s does not change and this is reflected in the higher slope of I–V curve near the open-circuit voltage.

On the other side, although the parameters calculated with the explicit equations of Section 2.2.2 ensure a better fit to the experimental data, such a procedure require knowing, apart from the short-circuit current and the open-circuit voltages, the exact position of the MPP for the environmental conditions under test. Unfortunately PV datasheets usually report detailed information only at STC, thus limiting the applicability of the method.

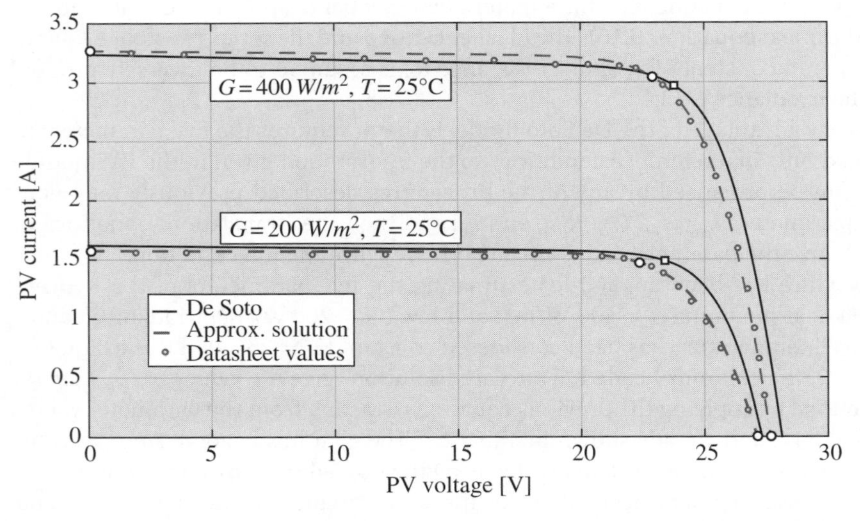

Figure 3.1 I–V curves of the Kyocera KC175GHT-2 panel obtained using De Soto's translated parameters and explicit equations.

3.2.3 Iterative Procedure proposed by Villalva et al.

The method proposed by Villalva et al. [14] is similar to the previous one except for the equation used for calculating of the saturation current. Now I_s is related to the open-circuit voltage while the temperature dependence is introduced by means of the α_v coefficient reported in the module datasheet:

$$I_{sc} = \frac{G}{G_0}\left[I_{sc0} + \alpha_i\left(T - T_0\right)\right] \tag{3.19}$$

$$I_{ph} = I_{sc}\frac{R_h + R_s}{R_h} \tag{3.20}$$

$$I_s = \frac{I_{sc0} + \alpha_i\left(T - T_0\right)}{e^{\left[\frac{V_{oc0}+\alpha_v(T-T_0)}{\eta V_t}\right]} - 1} \tag{3.21}$$

$$1 \leqslant \eta \leqslant 1.5 \tag{3.22}$$

$$R_{s,min} = 0 \tag{3.23}$$

$$R_{h,min} = \frac{V_{MPP}}{I_{sc0} - I_{MPP}} - \frac{V_{oc0} - V_{MPP}}{I_{MPP}} \tag{3.24}$$

The procedure starts by assigning to the five parameters the values obtained from (3.19)–(3.24). Subsequently the I_{ph}, I_s, R_s, and R_h parameters will be adjusted iteratively until the open-circuit voltage, the current and the voltage at MPP, and the short-circuit current fit with the corresponding reference values given by the datasheet or measurement.

The iteration stops when the maximum power (P_{MPP}) evaluated by means of (1.3) is very close to the value in the datasheet or obtained by measurement. Although this approach ensures good precision it suffers of some limitations:

- During the iterations, R_s is incremented in small steps ΔR_s starting from $R_s = 0$, so the convergence speed is affected by the choice of ΔR_s.
- The maximum power point on the P–V curve cannot be expressed explicitly with respect to the SDM parameters, so for each iteration it is necessary to evaluate the complete P–V curve by performing the multiplication $V_{pv} \cdot I_{pv}$ point by point, at the end finding the P_{MPP}. This process is time consuming and the accuracy depends on the resolution used to trace the P–V curve.
- No analytical equation has been provided for the ideality factor η. The authors suggest starting with a fixed value chosen arbitrarily, which is to be modified in order to improve the model fit. Unfortunately, no indication is given of how to modify such a parameter.

As shown by (3.19)–(3.24), the irradiance and temperature values affect directly only the calculation of I_{sc} and I_s; the other parameters are tuned by exploiting the STC information from the datasheet. Nevertheless, in order to fit the model to any other environmental condition, the Villalva procedure needs the values of I_{MPP} and V_{MPP} for the conditions under test, because the equations for calculating R_s and R_h depend directly on the actual MPP.

3.2.4 Modified PV Model proposed by Lo Brano et al.

Lo Brano et al. introduced a correction term in the SDM to improve its capability to reproduce I–V curves when the irradiance and temperature do not correspond to STC [15]:

$$I_{pv} = I_{ph} - I_s \left(e^{\frac{V_{pv}+R_s I_{pv}+\Psi \cdot I_{pv}}{\eta V_t}} - 1 \right) - \frac{V_{pv} + R_s I_{pv} + \Psi \cdot I_{pv}}{R_h} \tag{3.25}$$

$$\Psi = \frac{V_{MPP0} - V^*_{MPP}}{I^*_{MPP} \cdot (T^* - T_0)} \cdot (T - T_0) \tag{3.26}$$

The function Ψ is a thermal correction factor introduced to improve the SDM. V^*_{MPP} and I^*_{MPP} are the coordinates of the MPP at T^*. The V^*_{MPP} and I^*_{MPP} values might be provided by the manufacturer, but in general should be chosen by considering the maximum and minimum expected working temperature of the PV module. Some preliminary experimental measurements would therefore be performed to obtain an accurate Ψ value.

The other parameters in (3.25) are related to the environmental conditions by the following equations:

$$\eta = \eta_0; \quad R_s = \frac{R_{s0}}{\alpha_G}; \quad R_h = \frac{R_{h0}}{\alpha_G} \tag{3.27}$$

$$I_s(\alpha_G, T) = \frac{I_{ph}(\alpha_G, T) - V_{oc}(\alpha_G, T)/R_h}{\exp\left[\frac{V_{oc}(\alpha_G, T)}{\eta V_t}\right] - 1} \tag{3.28}$$

$$V_{oc}(\alpha_G, T) = V_{oc}(\alpha_G) + \alpha_V (T - T_0) \tag{3.29}$$

where $\alpha_G = G/G_0$ denotes the irradiance coefficient. The values of $I_{ph}(\alpha_G, T)$ might be evaluated with (3.4). The expression of $V_{oc}(\alpha_G)$, relating the open-circuit voltage with the solar irradiance, has been not provided. Lo Brano et al. stated that equations like (3.10) do not guarantee high accuracy and that it is preferable to use experimental values for $V_{oc}(\alpha_G)$ [15]. This is not a limitation on using the proposed method because $V_{oc}(\alpha_G, T)$ is used only for calculating the saturation current $I_s(\alpha_G, T)$.

An experimental campaign, performed using PV panels from many producers, has shown a regular dependence of $I_s(\alpha_G, T)$ on the solar irradiance in the range 200–1000 W/m^2, so the following expression might be used for obtaining the saturation current:

$$I_s(\alpha_G, T) = \exp\left[\left(\frac{\alpha_G - 0.2}{1 - 0.2} \right) \ln \frac{I_s(1, T)}{I_s(0.2, T) + \ln I_s(0.2, T)} \right] \tag{3.30}$$

Equation 3.30 is obtained by interpolation between the value of $I_s(1, T)$ calculated at $G = 1000$ W/m^2 ($\alpha_G = 1$) and the value $I_s(0.2, T)$ calculated at $G = 200$ W/m^2 ($\alpha_G = 0.2$). It is worth noting that, for calculating the values $I_s(1, T)$ and $I_s(0.2, T)$, Equation 3.28 is still required. This is not a limitation because many datasheets provide the values of V_{oc} for different irradiance conditions. From the electrical point of view, Equation 3.25 can be treated in the same way of the classical model of (1.3) by including the correction factor Ψ in series with the resistance R_s.

3.2.5 Translating Equations proposed by Marion et al.

The approach proposed by Marion et al. [2] is based on the assumption that, for a given irradiance and temperature condition (G, T), the PV voltage and current values (I_{pv}, V_{pv}) can be estimated by extrapolating the data of a reference I–V curve measured at the (G_0, T_0) conditions. The reference values are multiplied by two correction factors depending on the short-circuit and open-voltage values, as shown in the following equations:

$$I_{pv}(G,T) = I_{pv}(G_0,T_0)\frac{I_{sc}(G,T)}{I_{sc}(G_0,T_0)} \tag{3.31}$$

$$V_{pv}(G,T) = V_{pv}(G_0,T_0)\frac{V_{oc}(G,T)}{V_{oc}(G_0,T_0)} \tag{3.32}$$

$I_{sc}(G, T)$ and $V_{oc}(G, T)$ are determined with the following equations:

$$I_{sc}(G,T) = \frac{G}{G_0}I_{sc}(G_0,T_0)\left(1+\alpha_i(T-T_0)\right) \tag{3.33}$$

$$V_{oc}(G,T) = V_{oc}(G_0,T_0)\left(1+\alpha_v(T-T_0)\right)\left(1+\delta(T)\ln\left(\frac{G}{G_0}\right)\right) \tag{3.34}$$

In Marion et al. [2], a bilinear interpolation of four measured I–V curves are usually adopted as reference curves for extracting the temperature coefficients α_i, α_v and the coefficient $\delta(T)$ appearing in (3.34); in this way all the parameters are calculated by means of experimental data.

The translation procedure has been tested for a wide range of irradiances and PV module temperatures and for different PV module technologies. Root-mean-square errors in the range 1–5% have been obtained when the reconstructed data have been compared with the measured values of the maximum power, voltage and current at maximum power, short-circuit current, and open-circuit voltage. In this method it has been assumed that the fill-factor does not change, so part of the error in the MPP estimation is due to this approximation. The dependency of FF with respect to the irradiance and temperature conditions is strongly related to the PV panel under test, so the lower the FF variation the higher the accuracy assured by Marion's translating equations. Figure 3.2 shows the measured FF for several PV technologies during a sunny day [16]. Except for very low irradiance conditions, the FF variation is less than 10% for several PV technologies, which assures good results for Marion's translating equations.

It is worth noting that FF is an indicator of the quality of the PV cells. In Figure 3.2 PV panels having an almost constant FF are the ones made of hetero-junction cells. The HIT solution (heterojunction with intrinsic thin layer) at present represents the top technology for crystalline PV cells.

3.2.6 Modified Translational Equation proposed by Picault et al.

Picault et al. used the approach described in Section 3.2.5 to achieve a set of five equations for translating the parameters of the SDM for abitrary operating conditions [3]. Four measured reference curves are also needed to calculate accurately the temperature and irradiance coefficients. Nevertheless, as shown in the following, by performing some modifications to the Picault approach, the information extracted

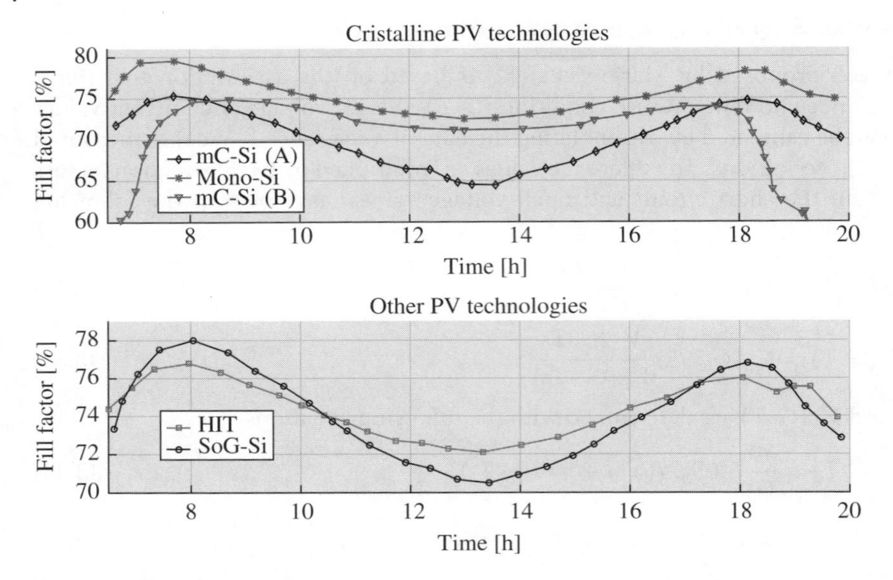

Figure 3.2 Measured fill factors of PV modules on a sunny day. Source: Yordanov et al. [16].

by the reference curves can be replaced with the STC parameters provided by the datasheet. Thus no preliminary experimental data are required to calculate the SDM parameters. Of course, the absence of experimental data is paid for with lower precision in reproducing the I–V curves in conditions that are far from the reference ones.

The reference values in the Marion approach can be replaced with the values calculated using the SDM model, so the following condition holds:

$$I_{pv}(G_0, T_0) = I_{pv0} = I_{ph0} - I_{s0}\left(e^{\frac{V_{pv0}+R_{s0}\cdot I_{pv0}}{\eta_0 V_t}} - 1\right) - \frac{V_{pv0} + R_{s0}\cdot I_{pv0}}{R_{h0}} \tag{3.35}$$

Moreover, by exploiting (3.31)–(3.32) it easy to verify the following equality:

$$V_{pv0} + R_{s0}\cdot I_{pv0} = \frac{V_{pv} + R_{s0}\frac{V_{oc}}{V_{oc0}}\frac{I_{sc0}}{I_{sc}}\cdot I_{pv}}{\frac{V_{oc}}{V_{oc0}}} \tag{3.36}$$

where V_{pv} and I_{pv} are the voltage and current corresponding to the generic environmental condition (G, T). Now, by substituting in (3.31) the variable $I_{pv}(G_0, T_0)$ with (3.35), and using (3.36), the $I_{pv}(G, T)$ current at any environmental condition is given by:

$$\begin{aligned} I_{pv}(G,T) &= I_{pv0}\frac{I_{sc}(G,T)}{I_{sc0}} \\ &= \frac{I_{sc}(G,T)}{I_{sc0}}\left(I_{ph0} - I_{s0}\left(e^{\frac{V_{pv}+R_{s0}\frac{V_{oc}}{V_{oc0}}\frac{I_{sc0}}{I_{sc}}\cdot I_{pv}}{\eta_0 V_t \frac{V_{oc}}{V_{oc0}}}} - 1\right) - \frac{V_{pv} + R_{s0}\frac{V_{oc}}{V_{oc0}}\frac{I_{sc0}}{I_{sc}}\cdot I_{pv}}{R_{h0}\frac{V_{oc}}{V_{oc0}}}\right) \end{aligned} \tag{3.37}$$

The values of the translated parameters I_{ph}, I_s, η, R_s, and R_h are calculated by collecting the terms in (3.37) as follows:

$$I_{ph} = I_{ph0}\frac{I_{sc}}{I_{sc0}}; \quad I_s = I_{s0}\frac{I_{sc}}{I_{sc0}}; \quad \eta = \eta_0\frac{V_{oc}}{V_{oc0}};$$
$$R_s = R_{s0}\frac{V_{oc}}{V_{oc0}}\frac{I_{sc0}}{I_{sc}}; \quad R_h = R_{h0}\frac{V_{oc}}{V_{oc0}}\frac{I_{sc0}}{I_{sc}} \tag{3.38}$$

The terms (I_{sc0}/I_{sc}) and (V_{oc}/V_{oc0}) are obtained using (3.4) and (3.10), so the five parameters are directly related to G and T:

$$I_{ph}(G,T) = I_{ph0}\frac{I_{sc}(G,T)}{I_{sc0}} = I_{ph0}\frac{G}{G0}\left(1+\alpha_i(T-T_0)\right) \tag{3.39}$$

$$I_s(G,T) = I_{s0}\frac{I_{sc}(G,T)}{I_{sc0}} = I_{s0}\frac{G}{G0}\left(1+\alpha_i(T-T_0)\right) \tag{3.40}$$

$$\eta(G,T) = \eta_0\frac{V_{oc}(G,T)}{V_{oc0}} = \eta_0\left[1+\alpha_v(T-T_0)+\frac{\eta_0 V_t}{V_{oc0}}\ln\left(\frac{G}{G_0}\right)\right] \tag{3.41}$$

$$R_s(G,T) = R_{s0}\frac{V_{oc}(G,T)}{V_{oc0}}\frac{I_{sc0}}{I_{sc}(G,T)} = R_{s0}\frac{1+\alpha_v(T-T_0)+\frac{\eta_0 V_t}{V_{oc0}}\ln\left(\frac{G}{G_0}\right)}{\frac{G}{G0}\left[1+\alpha_i(T-T_0)\right]} \tag{3.42}$$

$$R_h(G,T) = R_{h0}\frac{V_{oc}(G,T)}{V_{oc0}}\frac{I_{sc0}}{I_{sc}(G,T)} = R_{s0}\frac{1+\alpha_v(T-T_0)+\frac{\eta_0 V_t}{V_{oc0}}\ln\left(\frac{G}{G_0}\right)}{\frac{G}{G0}\left[1+\alpha_i(T-T_0)\right]} \tag{3.43}$$

The correction factors (I_{sc0}/I_{sc}) and (V_{oc}/V_{oc0}) can also be derived using the Marion equations (3.33) and (3.34); in this case the previous translating equations converge towards the formulas obtained by Picault [3].

The translating equations (3.39)–(3.43) have been applied for estimating the five parameters of the Kyocera KC175GHT-2 PV panel in medium ($G = 400\,\text{W/m}^2$) and low ($G = 200\,\text{W/m}^2$) irradiance conditions by starting from the corresponding values calculated at STC, as done with the De Soto equations. The parameters I_{ph0}, I_{s0}, η_0, R_{s0}, and R_{sh0}, are the ones in the the column headed "Exact solution" in Table 2.3.

In Figure 3.3 the I–V curves obtained with the five parameters calculated with the modified Picault equations and De Soto's approach have been compared with the experimental results extracted from the PV panel datasheet. The results highlight that, at least in the example under study, the modified Picault approach gives an overall good match. As concerns the MPP estimation, Table 3.3 reports the relative errors on the power,

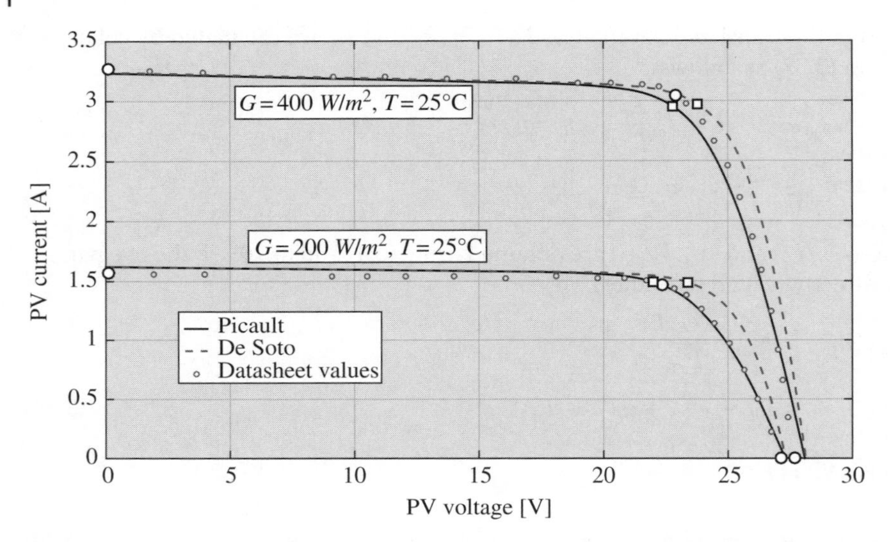

Figure 3.3 I–V curves of the Kyocera KC175GHT-2 panel obtained using modified Picault and De Soto translated parameters compared with the experimental data from the datasheet.

voltage and current values for the two methods. At the medium irradiance the Marion solution has a slightly higher error. The relative distance between the estimated and the real MPPs has been also calculated by normalizing the current and voltage values as follows:

$$MPP_{distance} = \sqrt{\left(\frac{V_{MPP,est}}{V_{oc}} - \frac{V_{MPP,meas}}{V_{oc}}\right)^2 + \left(\frac{I_{MPP,est}}{I_{sc}} - \frac{I_{MPP,meas}}{I_{sc}}\right)^2} \tag{3.44}$$

In Figure 3.3 the estimated MPPs are the squares and the real MPPs are marked with the circles. The figure shows that the modified Picault equations assure a better fit in terms of the MPP distance by reproducing more accurately the MPP in the I–V plane.

3.2.7 PV Electrical Model proposed by King et al.

King et al. [12] provide a detailed model of the photovoltaic module based on experimental data collected over many years at Sandia National Laboratories.[4] The model is formalized through a set of explicit equations characterizing five points of the I–V curve as a function of the irradiance and temperature at which the PV module works. The position of the five points is shown in Figure 3.4, while (3.45)–(3.51) highlight the dependency of such points on environmental conditions.

The equations can be used in real-time simulations where, by using the instantaneous values of the irradiance and temperature, the power predicted by King model can be compared with the measured PV power, providing information related to defects or malfunctioning of the PV system during its lifetime. The King model, if combined with

4 Sandia National Laboratories is a multi-program laboratory managed and operated by Sandia Corporation, a wholly owned subsidiary of Lockheed Martin Corporation, for the US Department of Energy's National Nuclear Security Administration.

Table 3.3 Errors in the MPP estimate for the Kyocera KC175GHT-2 panel.

	$G = 400\ \mathrm{W/m^2}$		$G = 200\ \mathrm{W/m^2}$	
	Marion	**De Soto**	**Marion**	**De Soto**
ΔP_{MPP} [%]	−3.49	1.3	−0.14	5.85
ΔV_{MPP} [%]	−0.43	3.93	−1.56	4.69
ΔI_{MPP} [%]	−3.07	−2.53	1.45	1.1
$MPP_{distance}$ [-]	0.028	0.04	0.018	0.04

one of the approaches described in the previous sections for calculating the five parameters of the PV non-linear equation, allows the entire I–V curve to be generated for any operating conditions.

$$I_{sc} = I_{sc0} \cdot \frac{G}{G_0} \left[1 + \alpha_{Isc}(T - T_0)\right] \tag{3.45}$$

$$\delta(T) = \frac{\eta k T}{q} \tag{3.46}$$

$$I_{MPP} = I_{MPP0} \left[c_0 \frac{G}{G_0} + c_1 \left(\frac{G}{G_0}\right)^2\right] \left[1 + \alpha_{I_{MPP}}(T - T_0)\right] \tag{3.47}$$

$$V_{oc} = V_{oc0} + N_s \delta(T) \ln\left(\frac{G}{G0}\right) + \alpha_{Voc}(T - T_0) \tag{3.48}$$

$$V_{MPP} = V_{MPP0} + N_s \left\{ c_2 \delta(T) \ln\left(\frac{G}{G_0}\right) + c_3 \left[\delta(T) \ln\left(\frac{G}{G_0}\right)\right]^2 \right\} + \alpha_{V_{MPP}}(T - T_0) \tag{3.49}$$

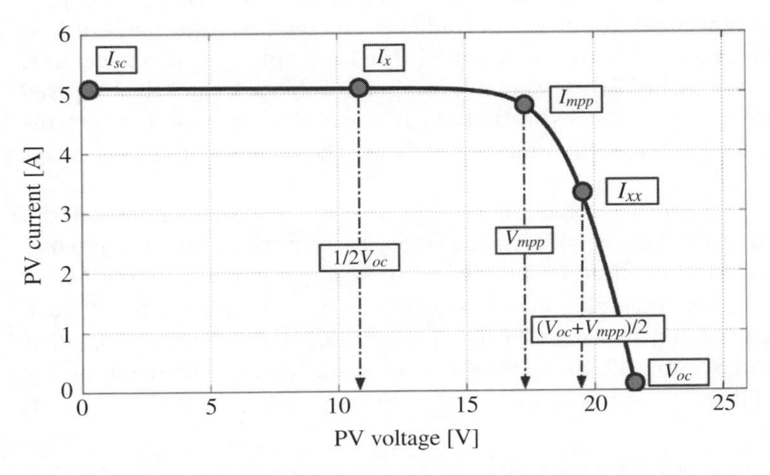

Figure 3.4 I–V curve showing the five points provided by the King's model.

$$I_{xx} = I_{xx0}\left[c_4\frac{G}{G_0} + c_5\left(\frac{G}{G_0}\right)^2\right]\left[1 + \frac{\alpha_{Isc} + \alpha_{I_{MPP}}}{2}(T - T_0)\right] \tag{3.50}$$

$$I_x = I_{x0}\left[c_6\frac{G}{G_0} + c_7\left(\frac{G}{G_0}\right)^2\right]\left[1 + \alpha_{I_{MPP}}(T - T_0)\right] \tag{3.51}$$

The King equations have been validated for flat-plate modules (all technologies) and for concentrator modules, with high versatility and accuracy. Unfortunately, for each PV panel under test, the model requires a set of parameters that must be evaluated experimentally. A database including all the performance parameters for many of the PV panels available in the market, is periodically updated on the Sandia website.[5]

The use of four separate temperature coefficients (α_{Isc}, $\alpha_{I_{MPP}}$, α_{Voc}, and $\alpha_{V_{MPP}}$) means that the King model applies to all photovoltaic technologies for a wide range of operating conditions. Tests have shown that the $\alpha_{V_{MPP}}$ and α_{Voc} coefficients are approximately similar, while the α_{Isc} and $\alpha_{I_{MPP}}$ coefficients show a significant difference in magnitude and in some cases also in the sign [17].

It is worth noting that in the module's datasheet only α_{Isc} and α_{Voc} are usually reported, so in some cases it is erroneously assumed that the temperature coefficient for V_{oc} is applicable for V_{MPP} and the temperature coefficient for I_{sc} is applicable for I_{MPP}. Fortunately, in the datasheet of the new generation of PV modules more data are available and the temperature coefficients $\alpha_{I_{MPP}}$ and $\alpha_{V_{MPP}}$ are explicitly reported [18, 19]. If not available from the module manufacturer, $\alpha_{I_{MPP}}$ and $\alpha_{V_{MPP}}$ must be measured in outdoor operating conditions [17]. For some commercial PV modules, all coefficients involved in the King model, are also available in the System Advisor Model (SAM) database.[6] The values of α_{Voc} and $\alpha_{V_{MPP}}$ can also vary with solar irradiance, especially for PV cells employed in concentrator photovoltaic systems. However, for flat-plate modules, constant values for the voltage temperature coefficients in (3.48) and (3.49) are generally adequate. Table 3.4 shows typical values of King's parameters for the most common PV technologies and it can be used when specific coefficients are not available.

The Kyocera KC175GHT-2 has been still used as reference because its datasheet provides many experimental curves for different irradiance and temperature conditions, so a wide range of operating conditions can be tested. The King model for this panel uses the coefficients reported in the "Polycrystalline" column of Table 3.4, since the PV panel under study does not appear in the SAM database. It is worth noting that, for the temperature coefficients α_{Voc} and α_{Isc}, the values provided in the manufacturer's datasheet have been used.

Figures 3.5 and 3.6 show the corresponding results: the square markers represent the five points described by the King equations. The circles correspond to the experimental data; the MPP for each test condition is marked with the larger circle. The continuous lines represent the curves obtained by using the SDM, where the parameters $\overline{\mathbf{P}}$ have been calculated by combining the King's subset equations (3.45)–(3.49) with the explicit equations reported in Section 2.2.2. The results show a satisfactory agreement at any environmental condition.

5 See http://www.sandia.gov/pv.

6 SAM software was produced by NREL in conjunction with Sandia through the US Department of Energy's Solar Energy Technologies Program. See https://sam.nrel.gov/.

Table 3.4 Parameters of the King model.

Coefficient	Silicon thin film	Single-crystalline	Polycrystalline	Triple-junction amorphous
$\alpha_{Isc\%}$	0.0916	0.0401	0.056	0.1263
$\alpha_{I_{MPP}\%}$	0.0358	−0.0390	−0.047	0.2034
$\alpha_{Voc\%}$	−0.4388	−0.3549	−0.3682	−0.4021
$\alpha_{V_{MPP}\%}$	−0.5629	−0.4560	−0.4830	−0.2976
η	1.357	1.026	1.025	3.09
c_0	0.9615	0.9995	1.0144	1.072
c_1	0.0368	0.0026	−0.0055	−0.098
c_2	0.2322	−0.5385	−0.3211	−1.8457
c_3	−9.4295	−21.4078	−30.2010	−5.1762
c_4	0.967	0.9980	0.9931	1.059
c_5	0.033	0.0020	0.0069	−0.059
c_6	1.12	1.159	1.104	1.188
c_7	−0.120	−0.159	−0.104	−0.188

Source: De Soto et al. [1]. The thermal voltage coefficients have been expressed in [%/K]. Additional data can be found in the System Advisor Model database at https://sam.nrel.gov/.

3.2.8 Using the King Equation for Estimating the SDM Parameter Drift

Experimental results, achieved for example by Khan et al. [20], show variations of all the SDM parameters that are not easily described with analytical functions. Therefore, for a more precise estimation of the PV operating conditions, the parameters' drift due to the environmental conditions should be determined with the support of experimental data. Nevertheless, by exploiting the King equations, which are customizable for any PV panel, the $\overline{\mathbf{P}}$ parameter drift due to the environmental conditions can be estimated with good accuracy and such values can be considered as a baseline. Figures 3.7 and 3.8 show the parameter variations for the BP Solar BP4175 and the Sanyo HIP-190BA2 PV panels. For these panels the King coefficients are available in the Sandia database, thus the translating equations can be applied straightforwardly. As confirmed by the values shown in Figure 3.8, the range and the way in which the parameters vary can be very different from panel to panel. For this reason, no general rules can be assumed for the $\overline{\mathbf{P}}$ parameter drift and simplified equations proposed in literature might be not always applicable.

On the other hand, the analysis shown in Figure 3.8 is based only on the I–V curve translation, so it is suitable for use both on the mathematical model as well as on the experimental data, and this information might be used for diagnostic purposes. Indeed, the $\overline{\mathbf{P}}$ parameters calculated in real time using experimental data might be compared with the corresponding values obtained with the King translating equations coupled with the SDM model. If the experimental data do not correspond with the model-based ones, within a pre-assigned level of confidence, an indication of the type of degradation is obtained. For example, a significant increase of the series

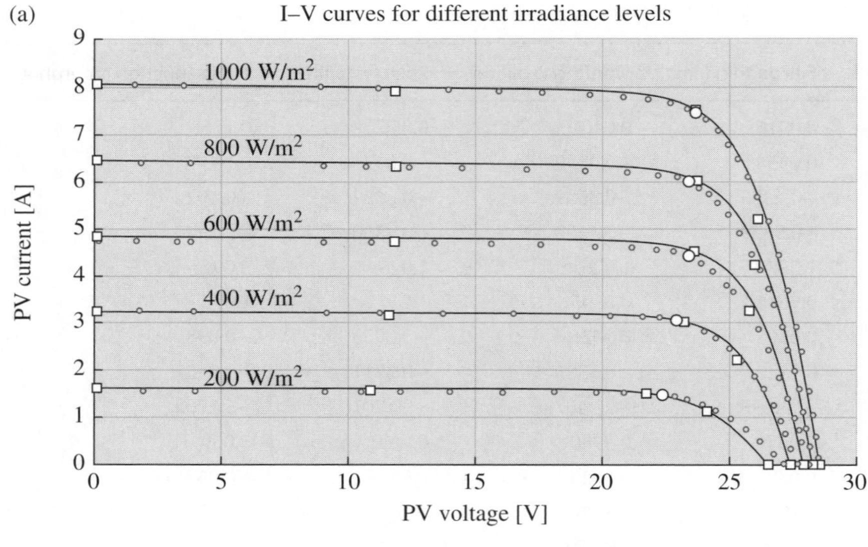

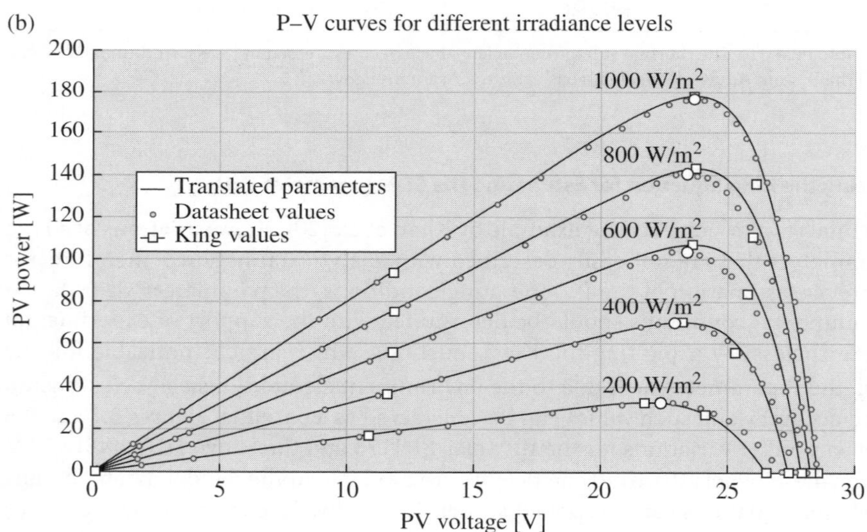

Figure 3.5 I–V curves of the Kyocera KC175GHT-2 panel obtained using the King translated parameters.

resistance can be associated with delamination effects or ribbon oxidation, while a reduction of the parallel resistance is associated with losses inside the PV cell, where the recombination phenomenon is increasing. More details about the analysis of the SDM parameters for diagnostic purposes have been shown in the literature [21–23], including some model-based indicators that have been introduced for real-time applications.

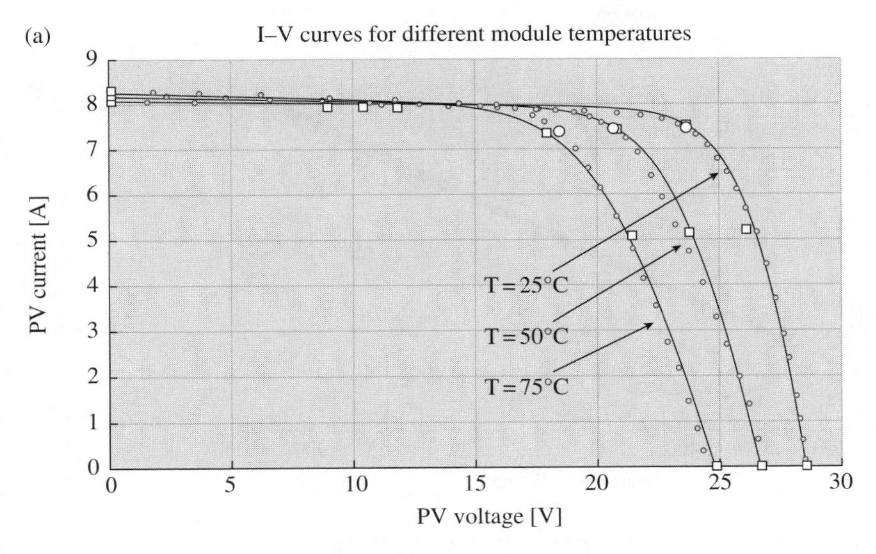

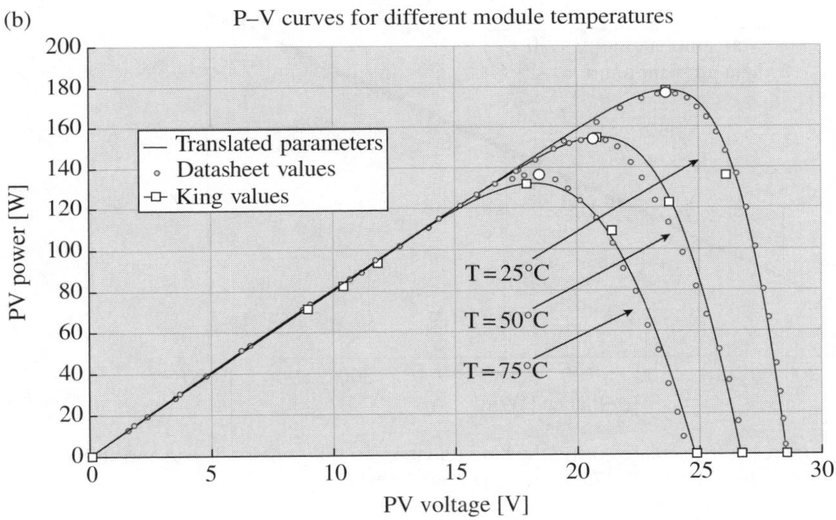

Figure 3.6 I–V curves of the Kyocera KC175GHT-2 panel obtained using the King translated parameters.

3.3 Simplified PV Models for Long-term Simulations

The cost per kilowatt-hour of the energy produced by new PV solar plants has decreased significantly, so their installation is becoming competitive with respect to non-renewable sources, even in the absence of incentives.[7] Of course, this is only

7 To encourage private investment in the production of electricity from renewable sources, many Countries, especially in the past, applied incentive policies by contributing to the cost of the PV installation or by promoting “feed-in tariff” solutions for selling the electricity produced by renewable sources.

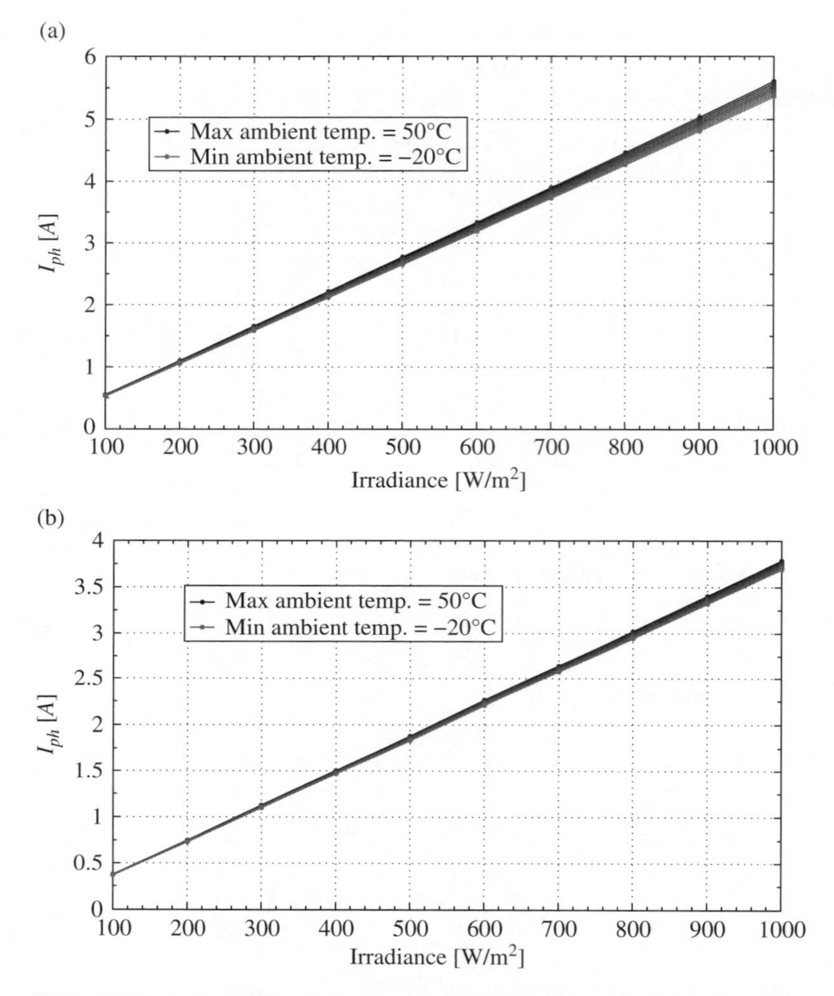

Figure 3.7 Parameter drift calculated using the King equation for different irradiance and ambient temperature levels: left, BP Solar BP4175; right, Sanyo HIP-190BA2.

possible if the PV plant works efficiently over its whole lifetime. Indeed, it has been estimated that, to reduce the investment payback time and give an attractive economic gain, the PV plant energy production must be more than 90% of the initial PV energy production over its lifetime [24]. Since the operating conditions of the PV fields depend on many factors, correlated in a complex way, the energy production is strongly variable and not easily predictable. For these reasons, in recent years there has been much interest in reliable methods for estimating the energy production of PV plants, and also in real-time monitoring and diagnostic functionality that will help reduce failures in PV sources and electronic equipment.

In the following, some models that are suitable for estimating PV energy productivity for long-term analysis will be outlined.

3.3.1 King Equations for Long-term Simulations

When information about the effective irradiance hitting the PV surface is not directly available, the model already introduced in Section 3.2.7 can be completed with the use of additional equations.[8] These equations allow use of irradiance information from meteorological websites or historical databases in calculating accurately the energy production of a PV field with an assigned orientation. Meteorological data usually refer to a

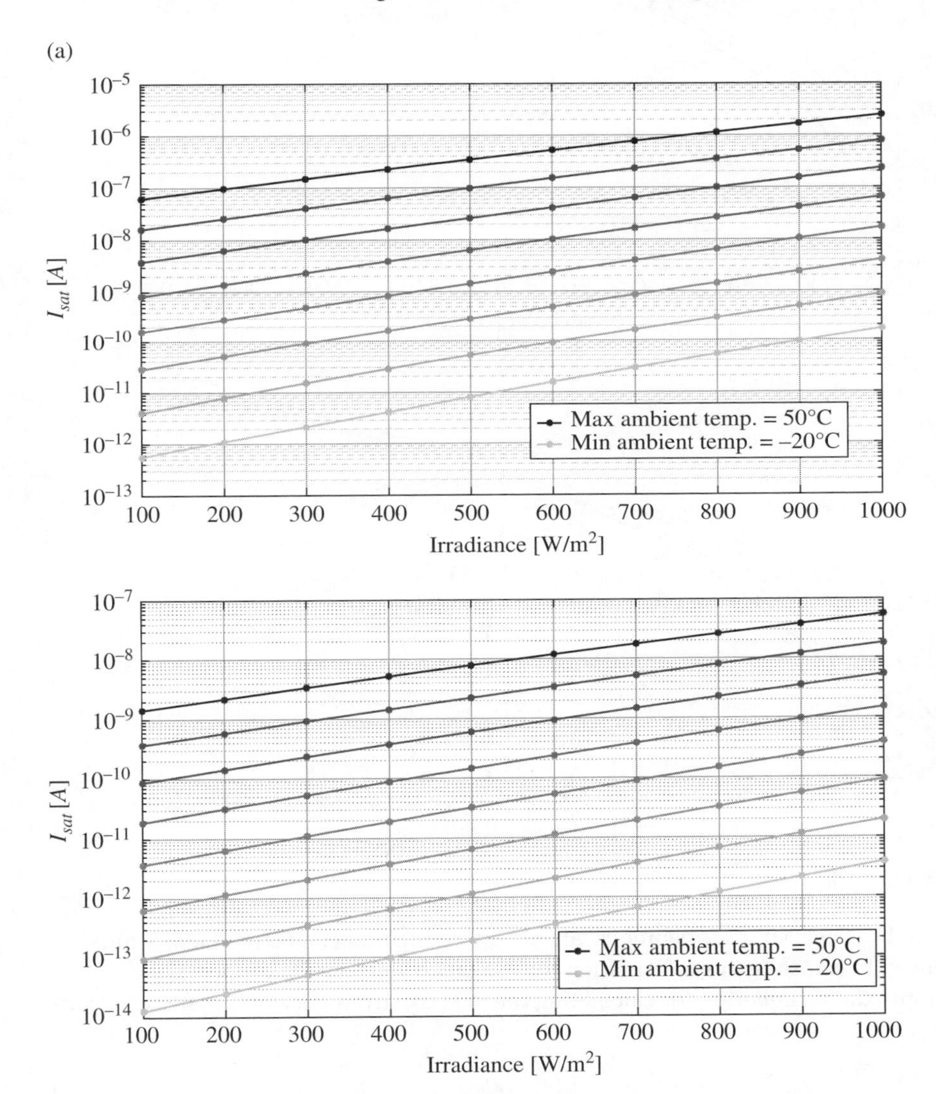

Figure 3.8 Parameter drift calculated using the King equation for different values of irradiance and ambient temperature: left column, BP Solar BP4175; right column, Sanyo HIP-190BA2. *(Continued)*

8 A direct measure of the effective irradiance can be obtained using a sacrificial PV cell as a sensor, which is placed on the same plane as the PV modules.

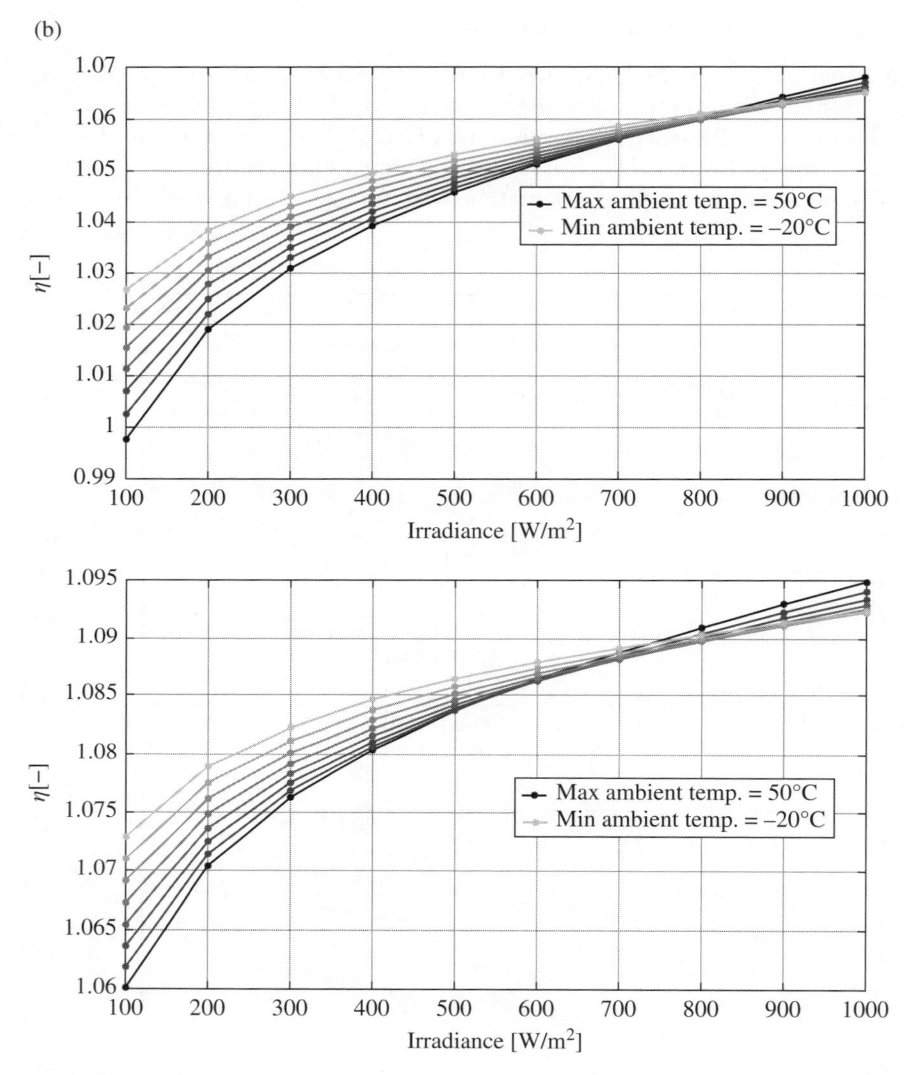

Figure 3.8 (Cont'd)

horizontal surface, so some geometrical and environmental correction factors must be introduced to translate this information for use on an abitrary PV surface.

An approximate equation for calculating the effective irradiance on a generic PV surface is [12]:

$$G \simeq f_1(AM) \cdot SF \cdot \left[G_b \cdot f_2(AOI) + f_d \cdot G_d\right] \tag{3.52}$$

where G_b and G_d are respectively the direct and diffuse component of solar irradiance on the PV module surface. In (3.52) the effect of the reflected component of the irradiance G_r has been included in G_d. For flat-plate PV modules, the coefficient f_d is set equal to 1; $f_d = 0$ is set for high concentration PV modules. The parameter SF – the so-called

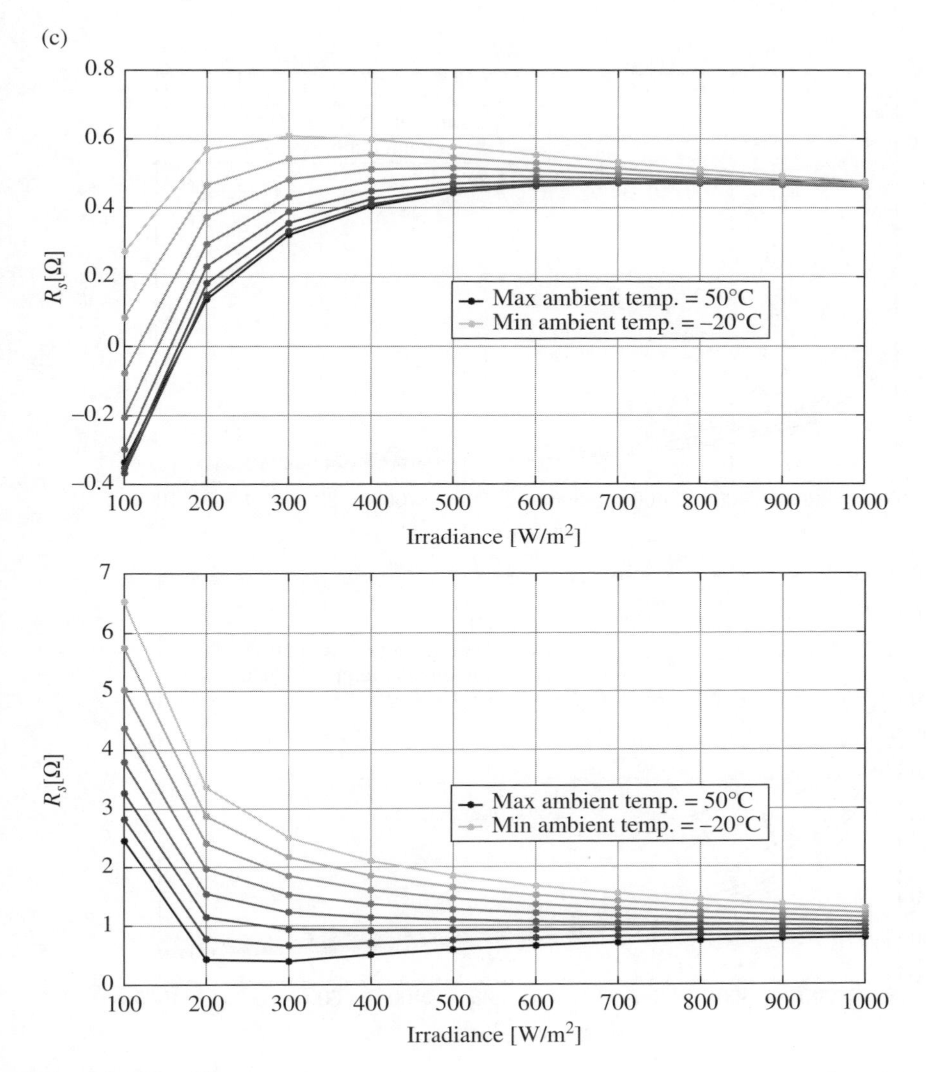

Figure 3.8 (Cont'd)

"soiling factor" – takes into account the losses due to dirt on the PV surface. Usually $SF > 0.95$, and $SF = 1$ corresponds to clean PV panels.

The isotropic sky model is widely used for calculating the direct and diffuse components of solar irradiance. It assumes that the diffuse radiation is uniformly distributed over the sky dome and that the reflection on the ground is diffuse, so that the solar irradiance components are given by:

$$G_b = G_{bh} \frac{\cos(AOI)}{\cos(\theta_z)} \tag{3.53}$$

$$G_d = G_{dh} \frac{1 + \cos(\beta)}{2} \tag{3.54}$$

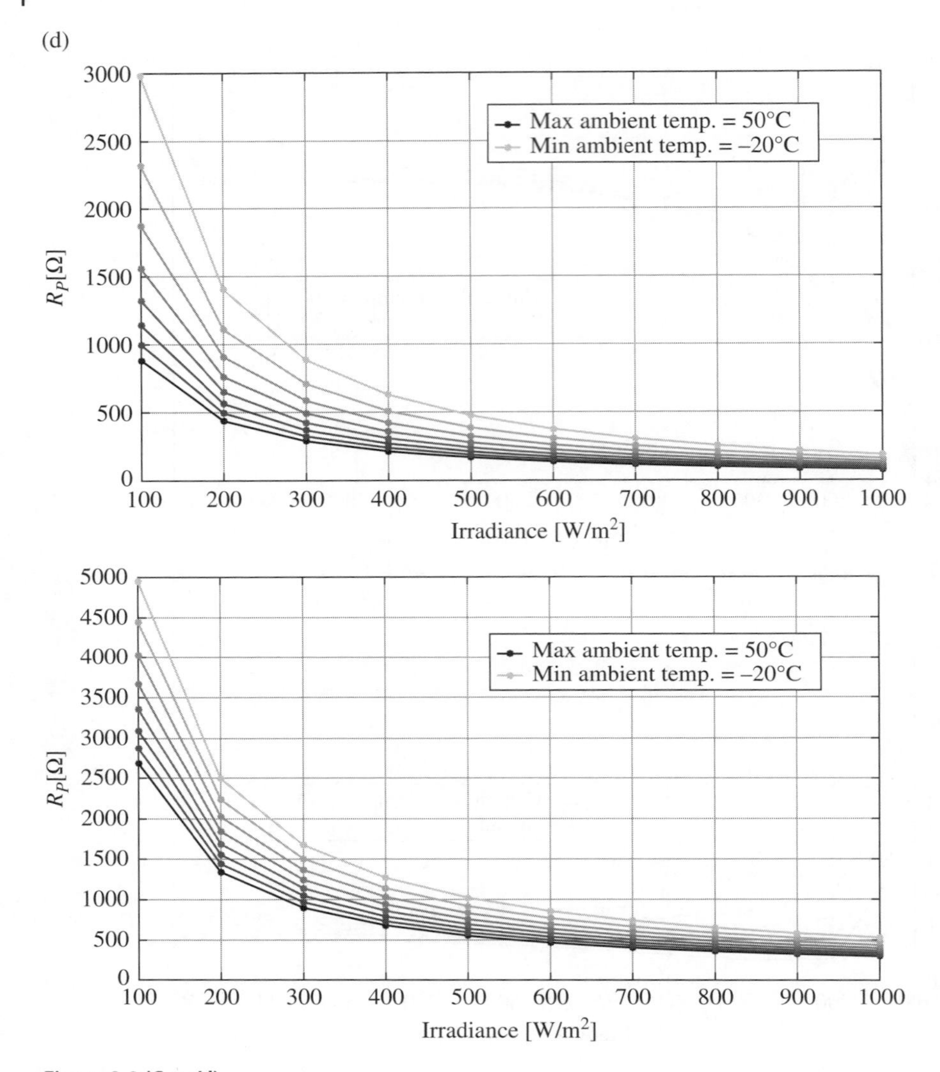

Figure 3.8 (Cont'd)

where G_{bh} and G_{dh} are respectively the direct and diffuse components of solar irradiance on the horizontal surface [25]. More accurate models for calculating G_b and G_d are reported in the literature [26].

The term $f_1(AM)$ in (3.52) is a function describing the influence of the air mass AM variation on G during the day. $f_2(AOI)$ is a function describing the optical influence of the solar angle of incidence AOI. Moreover, when the direct component of solar irradiance is not perpendicular to the module surface there are optical losses due to the increased reflections from the module materials that need to be accounted for.

The $f_1(AM)$ function can be approximated with a polynomial expression [27]:

$$f_1(AM) = \sum_{i=0}^{4} a_i \cdot AM^i \tag{3.55}$$

$$AM = \frac{1}{\cos(\theta_z) + 0.5057 \cdot (96.080 - \theta_z)^{-1.634}} \tag{3.56}$$

$$f_2(AOI) = \sum_{i=0}^{5} b_i \cdot AOI^i \tag{3.57}$$

$$AOI = \cos^{-1}\left[\cos(\theta_z)\cos(\beta) + \sin(\theta_z)\sin(\beta)\cos(\theta_A - \theta_{PV})\right] \tag{3.58}$$

where θ_A and θ_z are the solar azimuth and zenith angles, respectively. The parameters β and θ_{PV} are the tilt and azimuth angles of the PV array respectively.[9]

The coefficients a_i and b_i are estimated empirically by performing a fitting procedure between the above function and experimental data in clear-sky conditions. Table 3.5 shows the coefficients for several commercial PV technologies.

Once G has been estimated, (3.47) and (3.49) give directly the values of I_{MPP} and V_{MPP} for arbitrary environmental conditions, so they can be used for performing long-term energetic analyses based on the hourly, monthly, or annual meteorological data.

It is worth noting that the equations reported in this section might be used in combination with all the methods described in this book when it is necessary to have an estimation of the irradiance values on any arbitrary surface. A more detailed model for estimating the irradiance hitting the PV cells is described by Goss et al. [28]; this model

Table 3.5 Polynomial coefficients required for modeling spectral response.

Coefficient	Silicon thin film	Single-crystalline	Polycrystalline	Three-junction amorphous
a_0	0.938110	0.935823	0.918093	1.10044085
a_1	0.062191	0.054289	0.086257	−0.06142323
a_2	−0.015021	−0.008677	−0.024459	−0.00442732
a_3	0.001217	0.000527	0.002816	0.000631504
a_4	−0.000034	−0.000011	−0.000126	-1.9184×10^{-5}
b_0	0.998980	1.000341	0.998515	1.001845
b_1	−0.006098	−0.005557	−0.012122	−0.005648
b_2	8.117×10^{-4}	6.553×10^{-4}	1.440×10^{-3}	7.25×10^{-4}
b_3	-3.376×10^{-5}	-2.730×10^{-5}	-5.576×10^{-5}	-2.916×10^{-5}
b_4	5.647×10^{-7}	4.641×10^{-7}	8.779×10^{-7}	4.696×10^{-7}
b_5	-3.371×10^{-9}	-2.806×10^{-9}	-4.919×10^{-9}	-2.739×10^{-9}

Source: De Soto et al. [1]. Additional data can be found in the paper by King et al. [27].

9 The PV azimuth θ_{PV} indicates how the PV array surface is positioned with respect to due north. An array facing south has a PV azimuth of 180°. The PV tilt angle β is the angle the module surface makes with the horizontal plane.

generates an irradiance map of each cell of the PV array by considering both nearby and distant obstructions, thus accounting for shading conditions.

3.3.2 Performance Prediction Model based on the Fill Factor

Zhou et al used a semi-empirical equation for calculating the maximum power delivered by a PV field operating in uniform conditions [29]. This method combines the FF expression with two equations that describe the dependence of the short-circuit current and open-circuit voltage on the environmental conditions.

In the short-circuit current the temperature effect has been neglected because of its negligible impact, while the irradiance dependence has been considered as follows:

$$I_{sc} = I_{sc0}\left(\frac{G}{G_0}\right)^{\alpha} \tag{3.59}$$

The exponent α accounts for all the non-linear effects on the photo-induced current. It is calculated experimentally by measuring the short-circuit current for two irradiance conditions:

$$\alpha = \frac{\ln\left(\frac{I_{sc}}{I_{sc0}}\right)}{\ln\left(\frac{G}{G_0}\right)} \tag{3.60}$$

In the equation for calculating the open-circuit voltage V_{oc} at any given conditions, the temperature dependence is assumed to be no longer linear but instead expressed by the following equation:

$$V_{oc} = \frac{V_{oc0}}{1 + \beta \ln\left(\frac{G}{G_0}\right)}\left(\frac{T_0}{T}\right)^{\gamma} \tag{3.61}$$

where β is a dimensionless coefficient accounting for the PV module technology, γ is the exponential factor accounting for the all non-linear temperature-versus-voltage effects. The coefficient β is calculated by measuring the open-circuit voltage in two irradiance conditions and assuming a constant module temperature:

$$\beta = \frac{V_{oc0}/V_{oc1} - 1}{\ln(G_0/G1)} \tag{3.62}$$

The γ coefficient is calculated by measuring the open-circuit voltage at two different temperatures for the same irradiance value:

$$\gamma = \frac{\ln\left(\frac{V_{oc0}}{V_{oc1}}\right)}{\ln\left(\frac{T_1}{T_0}\right)} \tag{3.63}$$

The fill factor *FF* is a measure of the deviation of the real I–V characteristic from the ideal. The series and the shunt resistance, associated to the loss phenomena in the PV cell, reduce the fill factor and the delivered maximum power is reduced too. An empirical expression for the fill factor is given by:

$$FF = \frac{V_{MPP}I_{MPP}}{V_{oc}I_{sc}} \simeq FF_0\left(1 - \frac{R_s}{V_{oc}/I_{sc}}\right) \tag{3.64}$$

$$FF_0 = \frac{v_{oc} - \ln(v_{oc} + 0.72)}{1 + v_{oc}} \tag{3.65}$$

where FF_0 is the fill factor of the ideal PV module and v_{oc} is the normalized open-circuit voltage:

$$v_{oc} = \frac{V_{oc}}{\eta V_t} = \frac{V_{oc} q}{\eta k T} \tag{3.66}$$

η is the ideality factor (usually $1 < \eta < 2$).

The R_s and the η values appearing in (3.65) and (3.66) are still unknown. They can be calculated as follows:

$$R_s = \frac{V_{MPP}}{I_{MPP}} \cdot \frac{(I_{sc} - I_{MPP}) \cdot [V_{oc} + V_t \ln(1 - I_{MPP}/I_{sc})] - V_t \cdot I_{MPP}}{(I_{sc} - I_{MPP}) \cdot [V_{oc} + V_t \ln(1 - I_{MPP}/I_{sc})] + V_t \cdot I_{MPP}} \tag{3.67}$$

$$\eta|_{MPP} = \frac{V_{MPP} + I_{MPP} R_s}{V_{oc} + V_t \ln\left(\frac{I_{sc} - I_{MPP}}{I_{sc}}\right)} \tag{3.68}$$

Finally, the expression for calculating the power in MPP can be obtained straightforwardly by multiplying the fill factor expression by the terms I_{sc} and V_{oc}. Exploiting the corresponding expressions (3.59), (3.61), and (3.65), the P_{MPP} is given by:

$$\begin{aligned} P_{MPP} &= FF \cdot V_{oc} \cdot I_{sc} \\ &= \frac{v_{oc} - \ln(v_{oc} + 0.72)}{1 + v_{oc}} \left(1 - \frac{R_s}{V_{oc}/I_{sc}}\right) \\ &\times \frac{V_{oc0}}{1 + \beta \ln\left(\frac{G}{G_0}\right)} \left(\frac{T_0}{T}\right)^{\gamma} I_{sc0} \left(\frac{G}{G_0}\right)^{\alpha} \end{aligned} \tag{3.69}$$

Equation 3.69 allows the maximum deliverable power for any arbitrary environmental condition to be calculated. The five parameters (α, β, γ, R_s, and $\eta|_{MPP}$) are introduced to take into account all the non-linear effects of environmental factors on the PV module's performance, but unfortunately experimental data are required to tune the proposed model.

3.3.3 PV Modeling based on Artificial Neural Networks

In the recent literature, some applications of the artificial neural networks (ANN) to PV systems modeling have been presented. The ANN is an effective approach for describing a complex system that is strongly non-linear and depends on a large number of time-varying parameters, as is the case for PV generators. In fact, some parameters – such as the irradiation level and the operating temperature, but also some others related to the semiconducting material the cells are made of – may assume unpredictable values, which are also subject to drifts. The ANN structure describes a non-linear system on the basis of some weighted relationship between the given inputs (say, irradiance, temperature, or voltage) and the desired outputs (such as current). ANNs process the information in parallel through a large number of simple elements, which are the "neurons". All neurons are interconnected and each connection has a given weight. The neurons are distributed in layers and each neuron supplies an output through an activation function.

The ability of an ANN to learn from a training set, which might be made up from experimental data taken from the real PV generator working in different conditions, allows automatic tuning of the input–output relationship. Compared with the mathematical models described above, the ANN does not require a knowledge of the internal system parameters and involves less computational effort; it is a compact solution for such a multiple-variable problem.

Karatepe et al. studied the neural network shown in Figure 3.9 for estimating the $\overline{\mathbf{P}}$ parameters at arbitrary irradiance and temperature conditions [30]. A three-layer ANN was adopted: the input layer consists of a two-dimensional vector, incorporating the irradiation G and temperature T. The output vector is a five-dimensional vector comprising I_{ph}, I_s, η, R_s, and R_h. The hidden layer has 20 nodes. The hyperbolic tangent sigmoid function is used for the activation of the hidden layer (f_1 in Figure 3.9), while a pure linear function is chosen for the activation of the output layer (f_2 in Figure 3.9). The ANN learning phase allows the neuron weights ($w_{i,j}$, $w_{j,k}$) and bias (b_i^1,b_k^2) coefficients to be tuned using the back-propagation algorithm with the Levenberg–Marquardt optimization method. The latter is used to minimize the errors between the I–V curve calculated with the $\overline{\mathbf{P}}$ parameters estimated by the ANN and some reference curve. The King equations were used for reconstructing the I–V reference curves at the irradiance and temperature values adopted for the training set. The authors demonstrated very good accuracy for the curve reconstruction at a low irradiation, just where the classical approaches show larger inaccuracies with respect to experiments.

Almonacid et al. obtained a direct estimation of the electrical values of the PV panel using a multilayer perceptron (MLP) neural network [31]. In this case the MLP was trained using experimental measurements and it provided good results.

It is worth noting that in the ANN-based approaches the main elaboration burden is concentrated during the learning phase. The training process is very sensitive because

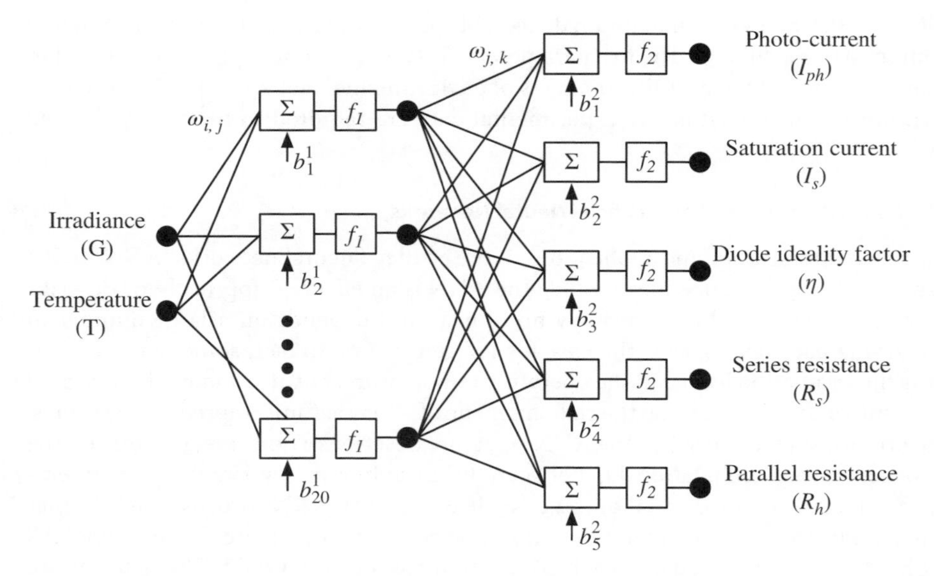

Figure 3.9 Neural network configuration for parameter estimation ($\overline{\mathbf{P}}$).

it requires the selection of a training set that is sufficiently representative of the module behavior. On the other hand, in order to avoid the ANN losing its generalization ability, the training procedure must be stopped before the error is reducing considerably. In order to balance such competing effects the ANN must be pre-tested on validation datasets, thus increasing the time taken for the learning process. Nevertheless, when the ANN is well tuned, it is a very useful tool for designers of PV systems since it can be used for quick estimations of the delivered power in long-term simulation windows by providing a suitable reference for the energy yield production of the PV system under study.

Piliougine et al. showed that it is important to take into account the spectral distribution of the incident light, especially at a low irradiance levels and for cell technologies that have a spectral response narrower than mono-crystalline silicon [32]. They included information about the spectral distribution of the light as a further input of the ANN. A non-random selection of the data used as the training set was implemented, with an improved performance of the network trained with the spectral information. The price to pay for this accurate modeling was in the amount of data required to represent the spectral information, and which has to be given to the ANN as an input, and also in the pre-processing of the training set, which is a time-consuming procedure. Piliougine's paper includes an up-to-date review of the most recent papers published in the field [32].

The unpredictable PV energy production makes the management of the power flow in the distribution network difficult. In this context, ANNs are also a valuable tool for PV energy forecasting. Ogliari et al. designed an ANN to give a one-day-ahead energy forecast on a hourly basis [33]. The ANN-based energy forecast can be improved if, at the end of each day, the meteorological data are combined with measured PV plant power production and used to activate a new ANN training process. In this way the ANN is tuned to give the energy forecast for the next day.

ANNs are also used for performing the MPPT function of PV plant operating in non-uniform conditions, where the MPP position depends on shading patterns, and the voltage at which it occurs may change within a large range. Syafaruddin et al. trained an ANN using many different partially shaded conditions to determine the PV voltage where the global MPP occurs [34]. The input signals for the ANN are the irradiance level G and the cell temperature T. The estimated MPP voltage V_{DC}^* and power P_{DC}^* were compared with the measured voltage V_{DC} and power P_{DC}. The predicted output voltage V_{DC}^* was also used as a reference signal for the MPPT controller, which was based on fuzzy logic. The latter is used to generate the required control signals for the power converter, as shown in Figure 3.10. The method has been experimentally validated, and the reliability and performance of the MPPT algorithm has been demonstrated to be superior to the conventional perturb-and-observe method in mismatched conditions.

3.4 Real-time Simulation of PV Arrays

The real-time simulation of PV systems is very useful in many applications, such as model-based control and rapid prototyping of circuits and systems used, again, for control purposes. To achieve this result, the implementation of the system models in digital electronic devices and embedded systems has to be very effective, both in terms

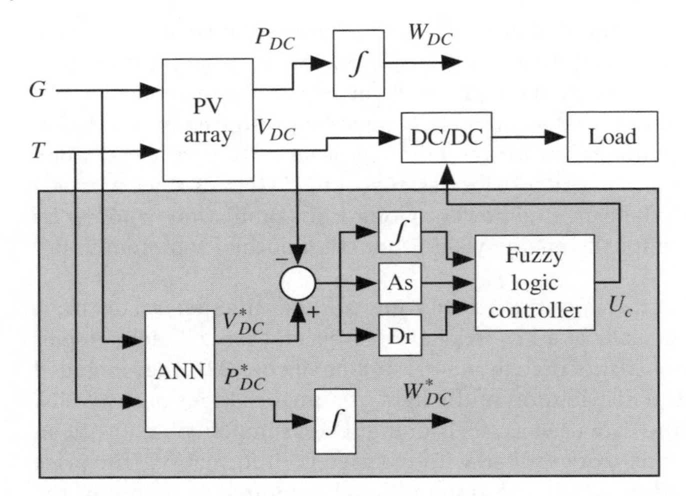

Figure 3.10 MPPT architecture for ANN with fuzzy logic controller.

of computation times and memory requirements. According to the recent literature, field programmable gate arrays (FPGAs) are the best candidates because of their computational features and the possibility of implementing parallel architectures. Mellit et al. [35] modeled the PV generator, as well as the power-processing system and a storage unit, through an ANN and implemented it in a commercial FPGA device. The implementation issues were discussed in relation to the limited resources of the device used and of the enhanced performance of the ANN achieved through the parallel computation features.

Embedded devices simulating PV arrays are frequently used for controlling power converters or supplies in order to emulate the static and dynamic behavior of a real system. Koutroulis et al. [36] discuss the implementation aspects of a PV simulator realized through an FPGA that controls a buck DC/DC converter switching at a maximum frequency of 300 kHz. The PV model used neglects the PV shunt resistance and reproduces the static I–V curves of a uniformly working PV generator, with a maximum discrepancy lower than 2%.

In their PV generator emulation, Gadelovits et al. use an off-the-shelf power supply, controlled using data corresponding to the desired I–V curve stored in a lookup table [37]. The real limitation of these kinds of approach is in the reproduction of the real variations of the I–V curve in the presence of a varying irradiance. Moreover, they are usually unable to reproduce the I–V curve of a PV array subjected to mismatching phenomena. These limitations are overcome by the technique presented by Di Piazza and Vitale [38], who use an SDM including only a series resistance, which gives good results in steady-state and dynamic conditions. These results are achieved by a proper design of the DC/DC switching converter. The emulation of the typical low-frequency disturbances produced by the single-phase grid connection in a real PV plant is also ensured.

3.4.1 Simplified Models including the Power Conversion Stage

In long-term simulations, both the PV source and the power processing system must be modeled with the objective of keeping the computational burden at a reasonable level.

Switching models, usually adopted for simulating electronic power converters, coupled with the non-linear model of the PV generator, result in slow and inefficient solutions because the size of the simulation step is directly related to the switching frequency of the power converters, so the higher the switching frequency, the smaller the simulation steps that are required. Moreover, switching models are usually implemented in circuital simulators where parasitic elements are also accounted for, so the simulation process becomes even more burdensome. For these reasons, simplified models describing the power processing system from a functional point of view are preferred when the simulations are being used to analyze system behavior over long timescales. Ropp and Gonzalez [39] show a Simulink model of a complete PV system.

Similarly Xiao et al. perform long-term simulations of grid-connected PV systems using average models of the power converters [40]. These models neglect the dynamics occurring in the switching period but preserve good precision in the low-frequency range, so are suitable for analyzing the effect of slow-varying environmental conditions on the converter stages and related controls by limiting the elaboration burden. In the following a brief description of this approach is given.

The analysis is focused on the scheme shown in Figure 3.11, a typical grid-connected PV system based on a two-stage topology with an intermediate DC link. In such an architecture, the DC/DC converter is devoted to maximizing the power extracted from the PV field, while the DC/AC stage performs the DC-link voltage regulation and the grid-tied functions, so that power quality and anti-islanding protection are ensured at the grid side. Because of the high capacitance of the DC link, the dynamic interaction of the DC/DC and DC/AC stages can be decoupled and studied independently.

From the photovoltaic point of view it is interesting to analyze the MPPT performance in different environmental conditions, so the analysis shown in the following is focused on the first stage only.

A DC/DC boost converter is used in the proposed example as shown in Figure 3.11. The switching model of the DC/DC power stage can be replaced by its averaged model, obtained using well-known averaging procedures that are introduced in Chapter 7, and which are discussed in depth elsewhere [41]. These methods allow the relationships between the average values of the current and voltage of the DC/DC stage and the control-variable duty cycle d to be modeled. For the circuit of Figure 3.11, the averaged model leads to the blocks shown in Figure 3.12, in which the state variables are V_{DC}, v_{pv},

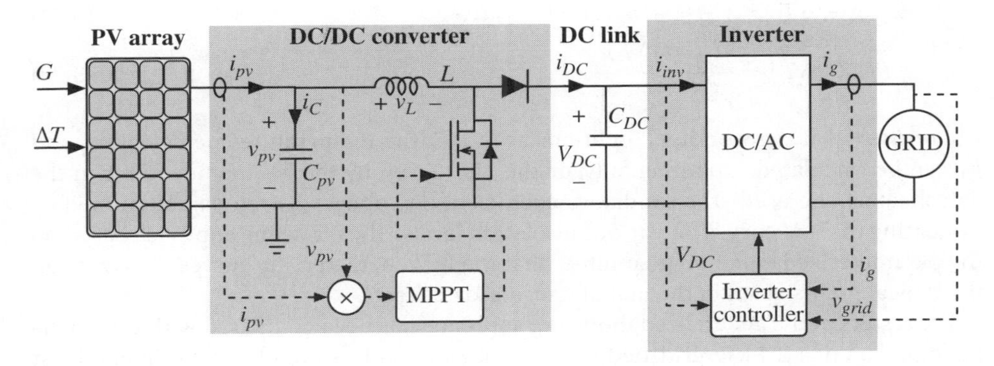

Figure 3.11 Grid-connected PV utility with DC/DC power converter and intermediate DC link.

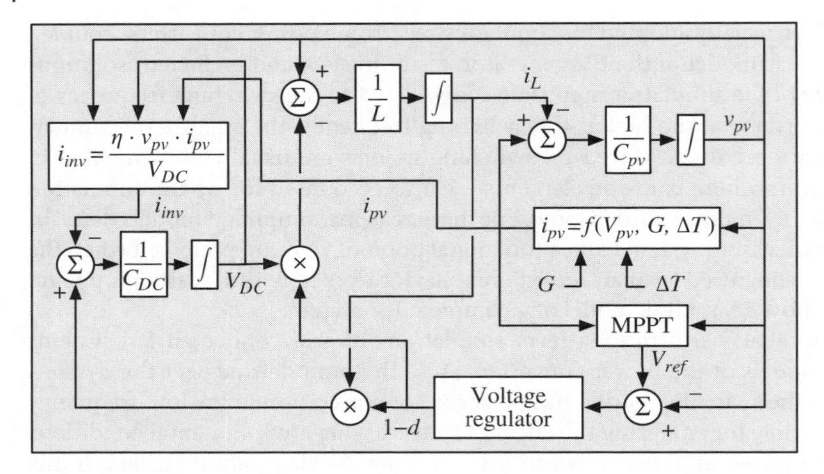

Figure 3.12 Averaged model of the PV array and the DC/DC power interface operating in continuous conduction mode.

and i_L. The power delivered into the grid depends on the converter efficiency and on the instantaneous power extracted by the PV array. Therefore, the instantaneous value of the current at the input of the inverter stage i_{inv} can be estimated as shown in the scheme. The MPPT block gives the reference voltage V_{ref} for the voltage loop aimed at tracking the optimal operating point of the PV source. The environmental parameters G and ΔT are the inputs used for calculating the instantaneous PV current i_{pv}.

The simple equation for the ideal SDM, shown in Section 1.4, is used for describing the PV source; the model has been parametrized directly with respect to the environmental condition as follows:

$$i_{pv} = I_{ph}(G, \Delta T) - I_s(\Delta G, \Delta T)\left[e^{\frac{qv_{pv}}{\eta k(T_{STC}+\Delta T)}} - 1\right] \tag{3.70}$$

The temperature and irradiance dependencies have been taken into account by means of the following equations:

$$I_{ph}(G, \Delta T) = \frac{G}{G_0} I_{sc0}(1 + \alpha_i \Delta T) \tag{3.71}$$

$$V_{oc}(\Delta G, \Delta T) = V_{oc0}(1 + \beta_T \Delta T)(1 + \gamma_E \Delta G) \tag{3.72}$$

$$I_s(\Delta G, \Delta T) = \frac{I_{ph}(G, \Delta T)}{e^{\frac{q \cdot V_{oc}(\Delta G, \Delta T)}{k\eta(T_{STC}+\Delta T)}} - 1} \tag{3.73}$$

where $\Delta G = 0$ if the working conditions are STC. The temperature coefficients α_i and β_T can be calculated experimentally, or the ones given by the PV manufacturer in the datasheet can be used. The irradiance coefficient on voltage, γ_E, can be determined by evaluating the I–V curves for various insolation levels. The modeling approach based on the parameterized equation guarantees that the I–V characteristic curves pass through the typical points given in the manufacturers' datasheets.

The results of the generalized modeling approach has been compared with the experimental data of a 2.4 kW grid-tied PV system. Figure 3.13 shows the experimental test performed in a single day. The solar power was available from 5:38am to 8:22pm. The

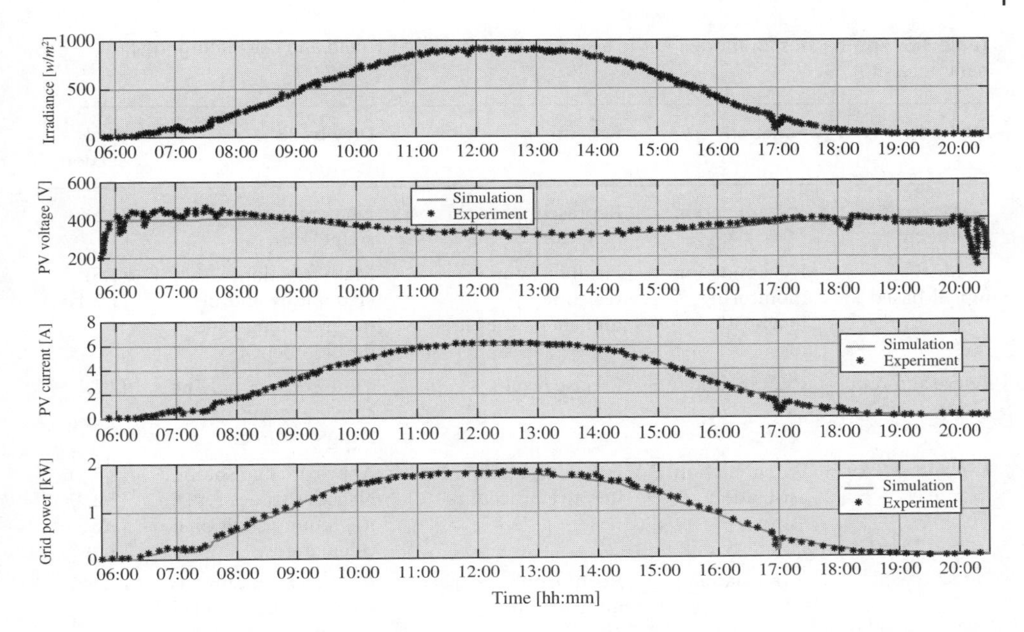

Figure 3.13 Comparison of measured and simulated waveforms from 5:38am to 8:22pm, 8 May 2010 in Vancouver, Canada.

simulated waveforms are compared with the experimental measurements of the PV voltage, PV current and the ac power injected into the grid; the waveforms almost overlap, confirming the good precision of the average model. Moreover, the average model does not require a tight simulation step, reducing considerably the computational burden and allowing long-term analysis of the whole PV system. Indeed, the simulation shown in Figure 3.13 takes only 20 s using a personal computer with single-CPU Intel i7–2.6 GHz and 16 GB RAM.

3.5 Summary of PV Models

The main advantages and drawbacks of the models discussed in Chapters 1–3 have been synthesized in Table 3.6 and are briefly compared in this summary section, so that the most appropriate one for a specific objective can be selected.

For simulating a PV field a fully circuital representation or model-based approaches might be used. Thanks to the availability of powerful simulators, detailed models can be implemented by means of computational block and/or programming code, allowing calculation of the I–V relationship using implicit or explicit expressions, as described in Section 1.5.

As concerns the identification of the $\overline{\mathbf{P}}$ parameters, the iterative procedures, described in Section 2.2.1 and the related literature, ensure precise results but have a higher computational burden and, in some cases, convergence problems can occur. Explicit formulas, such as the ones introduced in Section 2.2.2, give fast computations and ensure a good fit to experimental data, so are preferable for integration into a circuit simulator.

Table 3.6 Approaches for modeling the electrical behavior of a PV field and calculating the $\overline{\mathbf{P}}$ parameters.

Method	Function	Benefits	Drawbacks	Book section
SDM circuital approach	PV simulation	Represent physical behavior	No parameters adaptation	(1.5)
SDM model-based	PV simulation, monitoring, diagnostic	Flexibility and real-time implementation.	Math function elaboration and/or programming code development.	(1.5) (1.5.1)
Curve and root fitting	$\overline{\mathbf{P}}$ estimation	Precise results	Time consuming. The convergency is not always assured.	(2.2.1)
Explicit formulas	PV simulation, $\overline{\mathbf{P}}$ estimation	Fast computation. No iterative procedure	Approximated solution. Negative values for (R_s, R_h) require additional calculations.	(1.5.1) (2.2.2) (2.2.3) (2.3)
De Soto et al.	$\overline{\mathbf{P}}$ translation	Require only datasheet information.	R_s, R_h, η are assumed constant and calculated only at STC conditions	(3.2.2)
Villalva et al.	$\overline{\mathbf{P}}$ estimation and translation	Simple iterative procedure	Precise results only if V_{MPP} & I_{MPP} are known for the conditions under test.	(3.2.3)
Lo Brano et al.	PV modeling and $\overline{\mathbf{P}}$ translation	Simple to be tuned	Require experimental data	(3.2.4)
Picault et al.	$\overline{\mathbf{P}}$ translation	Explicit relation with respect to the environmental condition.	Assume a constant FF. Experimental data for tuning $\alpha, \beta, \delta(T)$ are required.	(3.2.6)
Modified Picault	$\overline{\mathbf{P}}$ translation	Require only datasheet information	Assume a constant FF. Datasheet parameters can be used for translating the SDM parameters.	(3.2.6)
Marion et al.	I–V translation	Simple	Assume a constant FF. Experimental data for tuning $\alpha_i, \alpha_v, \delta(T)$.	(3.2.5)
King et al.	I–V translation. Long-term energetic simulation	V_{MPP}, I_{MPP} as function of G and T. Additional estimated points (I_x, I_{xx}).	Many coefficients to be settled, requiring a large amount of data.	(3.2.7) (3.2.8) (3.3.1)
ANN	Long-term energetic simulation	Does not require a deep knowledge of the problem under investigation	Experimental data are mandatory	(3.3.3)
Simplified model	Long-term energetic simulation	Fast	Experimental data are mandatory	(3.3.2) (3.4.1)

Similar explicit formulas can be deduced for calculating the parameters involved in the model characterizing amorphous PV panels, as described and validated in Section 2.4.1.

The dependency of the $\overline{\mathbf{P}}$ parameters on irradiance and temperature conditions is widely treated in the literature. The De Soto and Villalva procedures [1, 14], described in Sections 3.2.2 and 3.2.3, respectively, are the most common ways to parametrize the SDM model. Both solutions assume that some parameters remain constant. This simplification might be acceptable if the scope of the model is to develop a PV array simulator/emulator for analyzing the electrical properties of the PV system where it is enough to reproduce with reasonable accuracy the trend of the I–V curve for any environmental condition. Such models can be easily integrated in circuital simulators for testing electronic circuits connected to the PV terminals or in MPPT algorithms; in such cases, high precision in the reproduction of the I–V curve is not required and it is recommended to use a simple model so as not to increase the computational burden by too much.

The modified Picault equations, described in Section 3.2.6, allow the $\overline{\mathbf{P}}$ parameters to be adapted for any arbitrary environmental conditions. With respect to other methods, the results are improved but constant FFs are assumed, which is not correct for some PV technologies or for some PV panels [16]. Thus this model is at best an approximation. Nevertheless, these methods do not require experimental data and are almost accurate even when far from the STC, provided they have been properly tuned.

The Marion and King equations are suitable for translating directly the PV voltage and current, so these methods can be used to estimate I_{sc}, V_{oc}, V_{MPP}, and I_{MPP} in any environmental conditions, after which the procedures for calculating the $\overline{\mathbf{P}}$ parameters can be used. However, additional data are required when these methods are used. For example, for the King equations, twelve coefficients must be provided. The fit is enhanced by preliminary measurements for the calculation of the coefficients.

Finally in Sections 3.3 and 3.4, some methods based on simplified equations and artificial neural networks have been described. The main advantage of such procedures over circuital and detailed model based approaches, is in the fast calculation of the energy production of the PV field in long-term simulations.

These models are based on the assumption that the PV field is under uniform conditions and that the operating point is the MPP. Unfortunately, as shown in the next chapters, such assumptions do not usually hold.

References

1 De Soto, W., Klein, S., and Beckman, W. (2006) Improvement and validation of a model for photovoltaic array performance. *Solar Energy*, **80** (1). 78–88.

2 Marion, B., Rummel, S., and Anderberg, A. (2004) Current-voltage curve translation by bilinear interpolation. *Progress in Photovoltaics: Research and Applications*, **12** (8), 593–607, doi:10.1002/pip.551.

3 Picault, D., Raison, B., Bacha, S., de la Casa, J., and Aguilera, J. (2010) Forecasting photovoltaic array power production subject to mismatch losses. *Solar Energy*, **84** (7), 1301–1309, doi:10.1016/j.solener.2010.04.009.

4 Liu, S. and Dougal, R. (2002) Dynamic multiphysics model for solar array. *Energy Conversion, IEEE Transactions on*, **17** (2), 285–294, doi:10.1109/TEC.2002.1009482.

5 Castaner, L. and Silvestre, S. (2002) *Modelling Photovoltaic Systems Using PSpice.* John Wiley.
6 Blaesser, G. and Rossi, E. (1988) Extrapolation of outdoor measurements of PV array I–V characteristics to standard test conditions. *Solar Cells*, **25** (2), 91–96, doi:10.1016/0379-6787(88)90014-2.
7 Marion, B. (2002) A method for modeling the current-voltage curve of a PV module for outdoor conditions. *Progress in Photovoltaics: Research and Applications*, **10** (3), 205–214, doi:10.1002/pip.403.
8 Davis, M.W., Fanney, A.H., and Dougherty, B.P. (2001) Prediction of building integrated photovoltaic cell temperatures. *Journal of Solar Energy Engineering*, **123** (3), 200–210, doi:10.1115/1.1385825.
9 Fuentes, M.K. (1987) A simplified thermal model for flat-plate photovoltaic arrays. *Sandia-Report, Unlimited Release.*
10 Skoplaki, E., Boudouvis, A., and Palyvos, J. (2008) A simple correlation for the operating temperature of photovoltaic modules of arbitrary mounting. *Solar Energy Materials and Solar Cells*, **92** (11), 1393–1402, doi:10.1016/j.solmat.2008.05.016.
11 Del Cueto, J. (2000) Model for the thermal characteristics of flat-plate photovoltaic modules deployed at fixed tilt, in *Photovoltaic Specialists Conference, 2000. Conference Record of the Twenty-Eighth IEEE*, Anchorage pp. 1441–1445, doi:10.1109/PVSC.2000.916164.
12 King, D.L., Boyson, W.E., and Kratochvill, J.A. (2004) Photovoltaic array performance model, in Sandia report SAND2004-3535, Sandia National Laboratories.
13 Skoplaki, E. and Palyvos, J. (2009) Operating temperature of photovoltaic modules: A survey of pertinent correlations. *Renewable Energy*, **34** (1), 23–29, doi:10.1016/j.renene.2008.04.009.
14 Villalva, M., Gazoli, J., and Filho, E. (2009) Comprehensive approach to modeling and simulation of photovoltaic arrays. *Power Electronics, IEEE Transactions on*, **24** (5), 1198–1208, doi:10.1109/TPEL.2009.2013862.
15 Brano, V.L., Orioli, A., Ciulla, G., and Di Gangi, A. (2010) An improved five-parameter model for photovoltaic modules. *Solar Energy Materials and Solar Cells*, **94** (8), 1358–1370, doi:10.1016/j.solmat.2010.04.003.
16 Yordanov, G., Midtgard, O.M., and Saetre, T. (2012) PV modules with variable ideality factors, in *Photovoltaic Specialists Conference (PVSC), 2012 38th IEEE*, pp. 002 362–002 367, doi:10.1109/PVSC.2012.6318073.
17 King, D., Kratochvil, J., and Boyson, W. (1997) Temperature coefficients for PV modules and arrays: measurement methods, difficulties, and results, in *Photovoltaic Specialists Conference, 1997, Conference Record of the Twenty-Sixth IEEE*, pp. 1183–1186, doi:10.1109/PVSC.1997.654300.
18 Yinglisolar. YGE-60 Cell Series web page: http://www.yinglisolar.com/assets/uploads/products.
19 Kyocera. KD 200-60 F Series web page: http://www.kyocerasolar.com/assets/001/5522.pdf.
20 Khan, F., Singh, S., and Husain, M. (2010) Effect of illumination intensity on cell parameters of a silicon solar cell. *Solar Energy Materials and Solar Cells*, **94** (9), 1473–1476, doi:10.1016/j.solmat.2010.03.018.
21 Bastidas-Rodriguez, J.D., Franco, E., Ramos-Paja, C.A., Petrone, G., and Spagnuolo, G. (2014) Model based indicators to quantify photovoltaic module degradation, in:

11th International Conference on Modeling and Simulation of Electric Machines, Converters and Systems, Valencia, Spain, 19–22 May 2014, pp. 49–54.

22 Bastidas-Rodriguez, J., Spagnuolo, G., Petrone, G., Ramos-Paja, C., and Franco, E. (2015) Model based degradation analysis of photovoltaic modules through series resistance estimation. *Industrial Electronics, IEEE Transactions on*, **62** (11), pp. 7256–7265.

23 Bastidas-Rodriguez, J., Franco, E., Petrone, G., Ramos-Paja, C., and Spagnuolo, G. (2015) Quantification of photovoltaic module degradation using model based indicators. *Mathematics and Computers in Simulation* doi:10.1016/j.matcom.2015.04.003.

24 Bazilian, M., Onyeji, I., Liebreich, M., MacGill, I., Chase, J., Shah, J., Gielen, D., Arent, D., Landfear, D., and Zhengrong, S. (2013) Re-considering the economics of photovoltaic power. *Renewable Energy*, **53**, 329–338, doi:10.1016/j.renene.2012.11.029.

25 Eicker, U. (2003) *Solar Technologies for Buildings*. John Wiley.

26 Loutzenhiser, P., Manz, H., Felsmann, C., Strachan, P., Frank, T., and Maxwell, G. (2007) Empirical validation of models to compute solar irradiance on inclined surfaces for building energy simulation. *Solar Energy*, **81** (2), 254–267, doi:10.1016/j.solener.2006.03.009.

27 King, D.L., Boyson, W.E., and Kratochvill, J.A. (1998) Field experience with a new performance characterization procedure for photovoltaic arrays, in *2nd World Conference and Exhibition on Photovoltaic Solar Energy Conversion*, Vienna, Austria.

28 Goss, B., Cole, I., Betts, T., and Gottschalg, R. (2014) Irradiance modelling for individual cells of shaded solar photovoltaic arrays. *Solar Energy*, **110**, 410–419, doi:10.1016/j.solener.2014.09.037.

29 Zhou, W., Yang, H., and Fang, Z. (2007) A novel model for photovoltaic array performance prediction. *Applied Energy*, **84** (12), 1187 –1198, doi:10.1016/j.apenergy.2007.04.006.

30 Karatepe, E., Boztepe, M., and Colak, M. (2006) Neural network based solar cell model. *Energy Conversion and Management*, **47** (9–10), 1159–1178, doi:10.1016/j.enconman.2005.07.007.

31 Almonacid, F., Rus, C., Hontoria, L., Fuentes, M., and Nofuentes, G. (2009) Characterisation of Si-crystalline PV modules by artificial neural networks. *Renewable Energy*, **34** (4), 941–949, doi:10.1016/j.renene.2008.06.010.

32 Piliougine, M., Elizondo, D., Mora-Lopez, L., and de Cardona, M.S. (2013) Photovoltaic module simulation by neural networks using solar spectral distribution. *Progress in Photovoltaics: Research and Applications*, **21** (5), 1222–1235, doi:10.1002/pip.2209.

33 Ogliari, E., Grimaccia, F., Leva, S., and Mussetta, M. (2013) Hybrid predictive models for accurate forecasting in PV systems. *Energies*, **6** (4), 1918–1929, doi:10.3390/en6041918. URL: http://www.mdpi.com/1996-1073/6/4/1918.

34 Syafaruddin, Karatepe, E., and Hiyama, T. (2009) Artificial neural network-polar coordinated fuzzy controller based maximum power point tracking control under partially shaded conditions. *Renewable Power Generation, IET*, **3** (2), 239–253, doi:10.1049/iet-rpg:20080065.

35 Mellit, A., Mekki, H., Messai, A., and Salhi, H. (2010) FPGA-based implementation of an intelligent simulator for stand-alone photovoltaic system. *Expert Systems with Applications*, **37** (8), 6036– 6051, doi:10.1016/j.eswa.2010.02.123.

36 Koutroulis, E., Kalaitzakis, K., and Tzitzilonis, V. (2009) Development of an FPGA-based system for real-time simulation of photovoltaic modules. *Microelectronics Journal*, **40** (7), 1094–1102, doi:10.1016/j.mejo.2008.05.014.
37 Gadelovits, S., Sitbon, M., and Kuperman, A. (2014) Rapid prototyping of a low-cost solar array simulator using an off-the-shelf dc power supply. *Power Electronics, IEEE Transactions on*, **29** (10), 5278–5284, doi:10.1109/TPEL.2013.2291837.
38 Piazza, M.C.D. and Vitale, G. (2010) Photovoltaic field emulation including dynamic and partial shadow conditions. *Applied Energy*, **87** (3), 814–823, doi:10.1016/j.apenergy.2009.09.036.
39 Ropp, M. and Gonzalez, S. (2009) Development of a Matlab/Simulink model of a single-phase grid-connected photovoltaic system. *Energy Conversion, IEEE Transactions on*, **24** (1), 195–202, doi:10.1109/TEC.2008.2003206.
40 Xiao, W., Edwin, F., Spagnuolo, G., and Jatskevich, J. (2013) Efficient approaches for modeling and simulating photovoltaic power systems. *Photovoltaics, IEEE Journal of*, **3** (1), 500–508, doi:10.1109/JPHOTOV.2012.2226435.
41 Erickson, R.W. and Maksimovic, D. (2007) *Fundamentals of Power Electronics*. Springer Science & Business Media.

4

PV Arrays in Non-homogeneous Conditions

The design of any PV system is based on an awareness of the phenomena that might lead a decrease in plant power production, in real conditions, over its whole lifetime. Some studies and evaluations have been done by researchers publishing in international journals, and by companies, especially those involved in the marketplace for electronics for distributed MPPT controls. These companies claim that such devices, well known as power optimizers and micro inverters, allow mitigation of the effects of mismatching phenomena that appear from time to time and that are usually neglected during the PV plant design phase.

Table 4.1 gives indicative values of the factors by which the array power production must be multiplied for phenomena which have to be taken into account in evaluating the plant power production.

Table 4.1 reveals that mismatching is one of the main sources of power reductions in a PV system's lifetime. The literature describes the closed loop existing between mismatching and cell aging, thus revealing the amplification of the detrimental effects of uneven aging of the cells.

In this chapter the main sources of mismatching are briefly analyzed and their effects on power production are described using data from the current literature, to which the reader is addressed for a more in-depth description of the operating conditions involved and for a wider discussion.

4.1 Mismatching Effects: Sources and Consequences

4.1.1 Manufacturing Tolerances

According to many studies, the differences in I–V curves of the modules used for building up a PV array are detrimental to its power production. Such differences, although small at the time of the plant construction, do not allow the modules to operate at their respective MPPs. For instance, data available in specialized magazines report that recently ninety thousands 300 W modules have been tested and their MPP voltages and currents varied in the ranges [34.5, 38.0] V and [7.89, 8.73] A respectively [1]. Binning the modules on a power basis means that heterogeneous values of voltages and currents will correspond to the MPPs of the modules, which are then connected in series and in parallel. Although many module manufacturers claim positive power tolerances, the worst module still acts as a bottleneck, thus limiting the power production of those

Photovoltaic Sources Modeling, First Edition. Giovanni Petrone, Carlos Andrés Ramos-Paja and Giovanni Spagnuolo.

Companion Website: www.wiley.com/go/petrone/Photovoltaic_Sources_Modeling

Table 4.1 Typical values of derating factors.

Category	Factor
Inverter and transformer	0.92
Soiling	0.95
Mismatch	0.98
DC wiring	0.98
AC wiring	0.99
Diodes and connections	0.995

having very high positive power tolerance. Indeed, the power-based binning might not mean that the MPP currents coincide.

Some manufacturers fix higher prices, about $0.02/W more, for modules that have peak power in a tighter range, but some studies suggest that this charge is unjustified. Indeed, by binning the polycrystalline panels on a current basis, the mismatch losses are about 0.6%, 0.16% and 0.01% for tolerances of 10%, 5% and 1% respectively. This is significantly less than the expected 1% [1]. According to the authors of this analysis, the reason is that the power–current curve of a PV module is almost flat across the MPP, so a reasonably small difference among the MPP currents of modules connected in series does not affect their power production significantly [1]. This conclusion holds for polycrystalline modules and even more so for CdTe ones, which are characterized by a smaller FF.

It is worth mentioning that manufacturers' data are obtained through flash testing. Unfortunately, this process is affected by an error because of the instrumentation used, which is even larger than 2%. There is also uncertainty about the temperature at which the modules' cells work during the flash test. Both these aspects affect the power range provided by the manufacturer, leading to an error in the tested power production of the module of up to 4–5%.

4.1.2 Aging

Aging is one of the main sources of mismatching. The module degradation is not uniform and recent studies have revealed that the power drop of a crystalline module is due to current and FF reductions, but not to significant changes in the voltage [2]. The PV module parameters are subjected to a change over time that is quantified by means of the ratio between the standard deviation and the mean value of the parameter value itself σ/μ. This figure increases from 1% to 10% for the module short-circuit current over 25 years of working time. It is greatly affected by the working temperature and humidity, which have an effect on delamination and discoloration of the materials used as encapsulants. These dependencies are analyzed in detail, by means of experimental data, in papers such as the one by Jordan et al. [2].

The mismatching among the short-circuit currents of the modules in the PV field, which increases with plant aging, determines a power loss, but it also increases the probability of hot spots, and thus permanent damage to the affected cells. In the literature it has been shown experimentally that this is the main cause of the σ / μ increase of the voltage value V_{MPP} at which the PV module MPP occurs. It has been shown experimentally that, in a cool and marine environment, the module P_{MPP} drops, on average, by 0.4% per year, in the first decade of operation, and by 1.4% per year in the second decade [3].

Aging is also the main cause of the so-called *snail trails*, which are small narrow dark lines and discolorations on the surface of the cell. An example can be seen in Figure 4.1a: the detail shown in Figure 4.1b reveals why this defect has taken the name it has. A recent experimental analysis of the effect of snail trails on PV modules shows a 40% module power production reduction [4].

4.1.3 Soiling and Snow

Dirt, pollution, snow, bird droppings and any other matter staying on the modules' surfaces reduces the number of photons reaching the cell surfaces. The effect of the accumulation of soil depends on the installation, on the site, and on the weather. A high frequency of rain would limit the power production because of the repeated lack of sun, but it would also keep the front surface of the PV modules clean. An example of visible soiling affecting modules is given in Figure 4.2.

In highly polluted and slightly rainy places, the derating factor due to soiling can reach a value of 25%. In some locations, also depending on the weather conditions, losses due to soil can reach up to the 70% of the total losses.

Snow does not slide off the modules, especially when they have been mounted with a small tilt angle, and this leads to a significant derating of the power production: an increase of the tilt angle from 23° to 40° can lead to a reduction from 70% to 40% of the derating factor. Snow can also have an indirect effect on the power production through the albedo, which is the reflecting power of a surface. Different positioning or orientation of the PV modules with respect to highly reflective surfaces – also an issue in the absence of snow – or a different amount of snow on some modules compared to others, can give rise to a mismatching effect due to the different irradiance level received by some modules. Some studies have analyzed the change in the albedo value due to snow. For instance, Andrews and Pearce studied the effect of snow on albedo compared to a grass surface [5]. The authors demonstrate that PV systems must also be accurately simulated in terms of reproduction of the climate and geographic location. Indeed, the use of an inaccurate value of the albedo leads to an underestimation of the PV array energy production of up to 10.5% for systems installed at a 90° tilt angle from the horizontal plane. High snow loads can also determine mechanical stresses on the panels, with deformation of the frame, breakage of the glass, or a slow creepage of the adhesive over a long time period.

4.1.4 Shadowing

Although in the neighborhood of the PV plant no sources of shading are present at the moment when the system is installed, in ground-mounted installations it must be accepted that, at least at the beginning and at the end of the day, some rows of modules or some even distant obstacles that were not present at the moment of the PV field installation, will produce a shadowing effect on a part of the PV array.

A practical example of what can occur at the sunset in a real installation is shown in Figures 4.3 and 4.4.

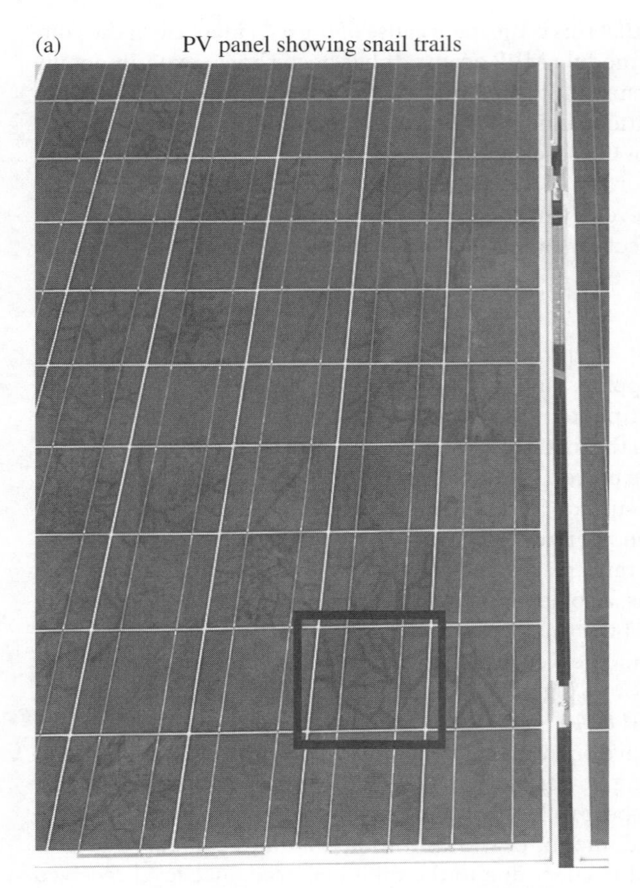

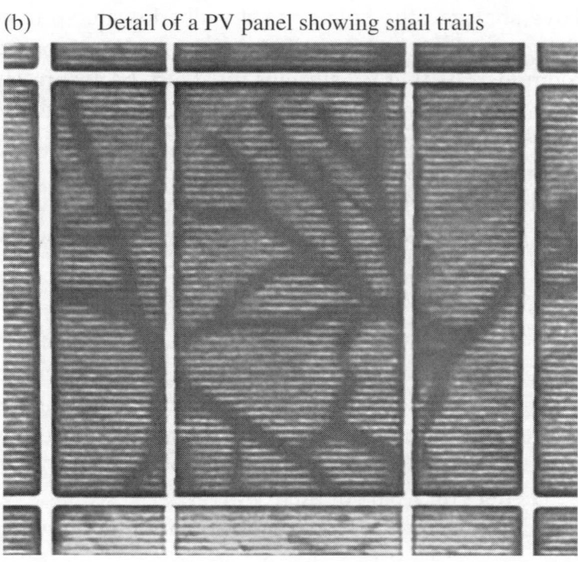

Figure 4.1 Snail trails affecting a PV panel. Source: SolarTech Lab of Politecnico di Milano, Italy.

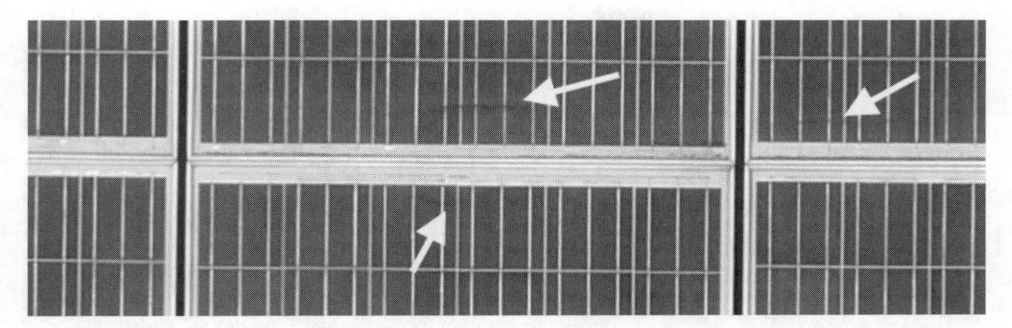

Figure 4.2 Dirt affecting a PV panel, visible as dark speckles in the bottom part of the upper side panels and in the upper part of the panels in the row below. Source: SolarTech Lab of Politecnico di Milano, Italy.

Figure 4.3 Shadowing of one string of PV panels by another.

Because of this, the distance between the rows of modules must be accurately calculated and the optimal value of the ground cover ratio (GCR), defined as the ratio between the PV field area and the total ground area, must be determined. In practice, the GCR is chosen so that a derate factor of 0.975 is obtained.

The shadowing has a non-linear effect on the power production of a PV string [6]. The shade impact factor (*SIF*) is used for quantifying this effect. It is defined as:

$$SIF = \left[1 - \frac{P_{shade}}{P_{array}}\right] \cdot \frac{A_{array}}{A_{shade}} \tag{4.1}$$

where P_{array} and A_{array} are the nominal power and the nominal area of the string, respectively. A_{shade} and P_{shade} are the shaded area and the PV power produced under the shadowing conditions considered, respectively. For instance, the *SIF* value would be close to

Figure 4.4 Shadowing due to an obstacle, such as a tree.

50 if the shadowing affecting the string causes an effect on the produced power that is 50 times greater than the relative extension of the shadow itself. *SIF* assumes a value close to unity if each cell is equipped with a bypass diode or if the array is uniformly shaded. Experimental results reveal that in a single-string grid-tied PV system, a shadow leads to a PV power reduction greater than 30 times its physical area [6]. For instance, for a flat-plate shadow falling on a single module it has been stated that the relationship between *SIF* and the module shade percentage is:

$$SIF = \frac{31.9}{\text{module shade percentage}} \tag{4.2}$$

This means that a module shade percentage greater than 15 lets the *SIF* almost fall below the value 2 [6].

4.1.5 Module Temperature

The modules' operating temperatures can be very mismatched because of, for example, the effect of wind or the type of surface the modules are mounted on. For instance, when it is integrated into a roof, depending on the availability of air gaps to improve the effect of air convection, the PV module can heat up, reaching a temperature on the back of close to 80°C [7]. Some examples have been presented in the technical literature: a 20°C difference between the hottest and the coldest module in a PV array mounted on a commercial rooftop in the USA has been recorded [8]. This temperature mismatch, on crystalline modules having a power thermal coefficient equal to 0.45%/°C, causes a 9% difference in the produced power between the hottest and the coldest modules in the array.

The performance of crystalline silicon modules is much more sensitive to temperature than are amorphous silicon ones. This makes the effect of temperature mismatching much more dramatic in the former case [7].

4.2 Bypass Diode Failure

As will be analyzed in detail in Chapters 5 and 6, bypass diodes are key components when the PV array is forced to work in mismatched conditions because of external (e.g. shadowing) or internal (e.g. manufacturing tolerances) factors. They allow more power to be extracted from the PV array and to limit the probability of permanent damage to the cells. In some cases, modules are shipped from the factory with some bypass diode failures.

The bypass diodes are mounted in the junction box, which is small, closed and placed at the rear of the PV module. This affects the working temperature of diodes significantly and has a great impact on the bypass diodes' performance, so it is important that the junction box be designed properly.

The diode must enter into conduction when even only one cell is shadowed. Thus, in a string of n PV cells in series, the condition for having the bypass diode turn ON when one cell works at a negative voltage is:

$$n < \frac{V_{shadowed} - V_{bypass}}{V_{non-shadowed}} + 1 \tag{4.3}$$

where $V_{shadowed}$ is the voltage at which the shadowed cell works, V_{bypass} is the forward voltage at which the bypass diode turns ON and $V_{non-shadowed}$ is the forward voltage at which each irradiated cell works. These variables are shown in Figure 4.5. By assuming that $V_{bypass} = 0.5$ V, $V_{non-shadowed} = 0.5$ V, and $V_{shadowed} = 12$ V mean that $n < 24$, which is the threshold usually adopted by the manufacturers. It should be pointed out that

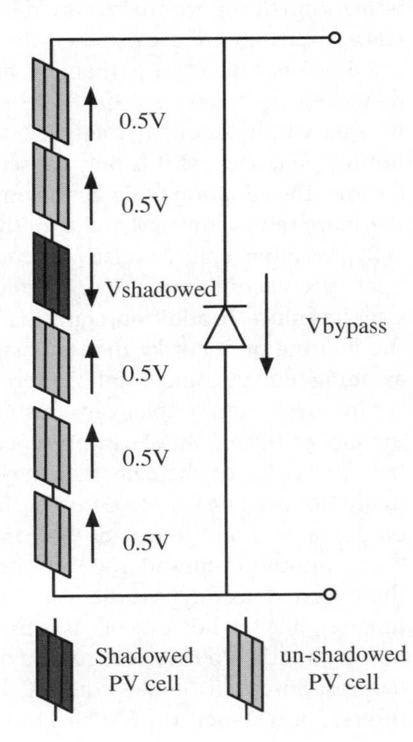

Figure 4.5 PV string with one cell shadowed.

$V_{shadowed}$ is the PV cell breakdown voltage, whose value ranges between 12 V and 20 V for polycrystalline cells and goes up to 30 V for monocrystalline ones.

Some producers are trying to address this problem by reducing the breakdown voltage of their cells. This is the case of SunPower [9], which produces standard modules having a cell-breakdown voltage of 17 V with an MPP STC voltage of 29.5 V. The SunPower X21 345-W module has a cell breakdown voltage of 2.5 V and an MPP STC voltage equal to 57.3 V.

The mostly widely used diode is the Schottky type because of its low forward voltage. Nevertheless, this parameter cannot be the only one for choosing the right diode for the bypass application. Indeed, for a Schottky diode, a lower forward voltage means a higher leakage current, which means a higher loss when the bypass diode is switched OFF. The reverse losses, which might affect the module efficiency by 0.05%, increase with the junction temperature [10]. It has been estimated that there is a doubling of the leakage current for each temperature increase of 10°C. Unfortunately, the leakage current is directly related to the thermal runaway risk, which occurs when the bypass diode is subjected to an ON-OFF-ON sequence. When the bypass diode is turned ON, because the PV string it protects is partially shadowed, its temperature increases, up to 200°C. As soon as the shadow disappears, it turns OFF, so it is reverse biased. Because of the high temperature it has reached in the forward mode and due to the lack of convection effects in the junction box where the bypass diode is located, the leakage current increases abruptly [10]. The power losses in the reverse mode are the product of this leakage current multiplied by the reverse voltage. It is worth noting that at 150°C the diode leakage current is above 100 mA. The occurrence of the thermal runaway phenomenon is proportional to the diode reverse voltage. If the power losses due to the leakage current are greater than those in forward mode, thermal runaway occurs and the diode is subject to permanent damage. It is evident that the main reason for this damage is the lack of any air flow in the junction box. Indeed, laboratory experiments in oven with systems that equalize the chamber temperature, the diode failure might not happen, because it is only caused by the total absence of any internal junction box airflow. The adoption of diodes having lower heat dissipation during the turn ON would therefore reduce the probability of thermal runaway.

Bypass diode failure is also induced by lightning strikes [11]: the diode remains in the open-circuit condition, thus potentially leading to hot spots if partial shadowing occurs after its failure. In addition, static high-voltage discharges and mechanical stresses affect the lifetime of Schottky diodes very significantly, so they must be handled carefully, avoiding human contact unless there is grounding.

Moreover, even simpler causes, such as soldering disconnections, can cause the turning ON of bypass diodes in presence of partial shadowing to be missed, with serious hot-spot risks for the cells that they should protect. Indeed, the bypass diode failure might not even be easy to detect by inspection. Burn marks indicating defective bypass diodes appear on the rear of the panel, but not always. Indeed, an experimental study based on one thousand 180 Wp photovoltaic modules of a single type revealed that there were defect bypass diodes with burn marks only in the 3% of cases [12]. Unfortunately, in a further 44% of cases the defect in the bypass diodes was not accompanied by burn marks. In total, 47% of the diodes suffered evident and especially non-evident failures. So-called "lossless diodes" have been recently introduced by some manufacturers. For instance, the SM74611 Smart Bypass Diode of Texas Instruments ensures a typical forward power dissipation of 208 mW, which is significantly lower than the 3.2 W of conventional Schottky diodes. ST Microelectronics produces a Cool Bypass Switch SPV1001 with a low forward voltage drop and reverse leakage current. These results are achieved by using a power MOSFET, which charges a capacitor during the OFF time, and using the stored charge to drive its own gate during the ON time.

Examples of active bypass devices have been recently outlined in the literature too [13].

4.3 Hot Spots and Bypass Diodes

Hot spots appear in cells when they are forced to work with a high forward current, imposed by the other highly irradiated cells connected in series, and a high reverse voltage, which is only limited by the cell PN junction breakdown voltage. The cell then absorbs a significant amount of power, instead of producing it. When this power exceeds the critical power dissipation of the cell, which is dependent on the cell working conditions (such as ambient temperature or convective effects) and on material structure, the cell that is sinking too much power can be damaged permanently. Figures 4.6 and 4.7 show the way in which the hot spot becomes evident on the panel surface and on the cell-temperature distribution detected using a thermo-camera.

A shadow affecting a small portion of the whole area of the PV array, thus appearing to be negligible, can lead to a local cell-temperature increase. An example is given in Figure 4.8, which shows the shadow produced by two medium-voltage power cables on some PV panels (see Figure 4.8a). In Figure 4.8b, thermography reveals that the temperature of the two hot spots, HS1 and HS2, is 55% higher than the temperature of the coldest points, CS1 and CS2.

A detailed simulation analysis of thin-film modules illustrates the problems related to the partial shadowing and the cases in which the bypass diode becomes ineffective [16].

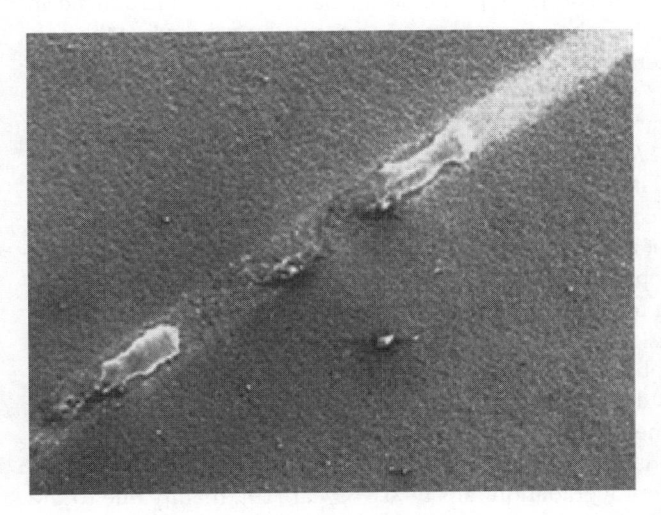

Figure 4.6 Solder melt at hot-spot site. Source: Simon and Meyer (2010) [14].

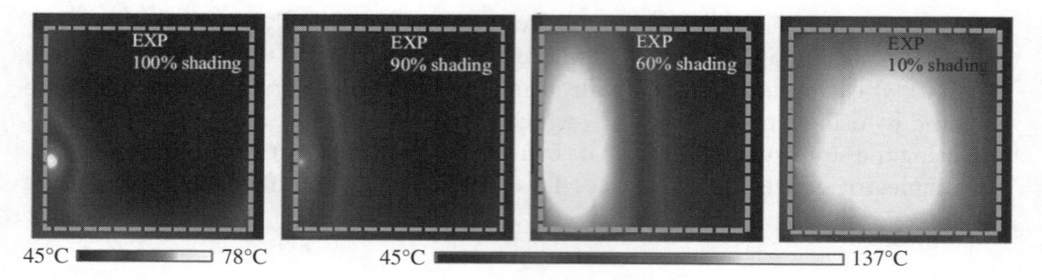

Figure 4.7 Hot spot temperature and shadowing area. Source: Geisemeyer et al. (2014) [15].

The authors show that small shadows are the most dangerous in terms of the stresses the cell is subjected to, both for the shadowed and for the un-shadowed portions.

Some recent studies have also shown the limitations of the approach based on bypass diodes for crystalline modules [17].

4.4 Effect of Aging Failures and Malfunctioning on the PV Energy Yield

The performance of a PV plant is usually evaluated by means of the so-called performance ratio (PR). It is defined as the ratio of measured PV energy and the theoretical energy yield:

$$PR = \frac{E_{ac}}{\frac{P_{STC}}{G_{STC}} \int_{t_0}^{t_1} G dt} \tag{4.4}$$

where E_{ac} is the net electrical energy produced by the PV system during the period $[t_0; t_1]$. P_{STC} is the rated power of the PV generator at STC, G_{STC} is the global solar irradiance under STC. G is the time-varying global solar irradiance received by the PV generator during the reference period $[t_0; t_1]$. The performance ratio is introduced in the IEC 61724 standard, and it is an harmonized metric suitable for comparing the performance of different PV systems over a specified time period, for example monthly or yearly estimation.

The PR is a cumulative measure of whole-system losses, aging, malfunctioning and failures, so the analysis of the PR's variation over a time window is also used for diagnostic purposes. Indeed, if the PR is going down below a reference value (PR_{ref}), the system is operating under abnormal conditions or that there has been a failure. The selection of such a reference value is not a trivial task. Indeed, if PR_{ref} is too high (an indicator of a high-quality PV system) the probability that $PR < PR_{ref}$, giving a false failure identification, increases thus alerting maintenance teams needlessly, is high. Conversely, a low reference value leads to the risk of a delayed or missed alert of the PV system malfunctioning because it is highly probable that $PR > PR_{ref}$, even in presence of partial failures. Usually the reference value is defined on the basis of analytical models and long-term historical data. Several documents highlight that the PR value is strongly variable, and in normal operating conditions it spans the range 70–90% [18]. It has been estimated that, on average, the PV system degradation falls by 0.8%/year [19], mainly due to the intrinsic aging of PV panels. Figure 4.9 shows a statistical distribution of PV plants from

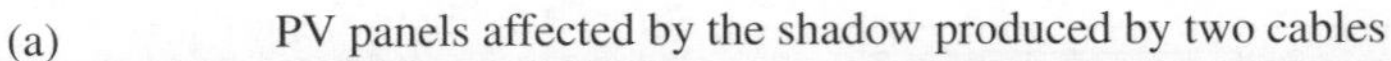

Figure 4.8 Cable shadowing affecting some PV panels. Source: SolarTech Lab of Politecnico di Milano, Italy.

this analysis [19]. The basic *PR* definition has been further refined in order to take into account the thermal drift and the converter efficiency variation. The statistical distributions of the *PR* indicator and the analyses of some correction factors are found in the literature [18, 20].

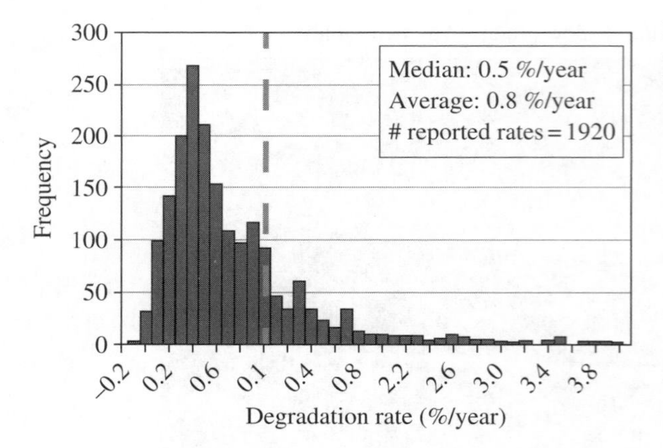

Figure 4.9 Photovoltaic panel degradation rate. Source: Jordan and Kurtz (2013) [19].

Different from aging effects, the estimation of the energy losses due to failures is much more complicated because, it is mandatory to identify only the PV plant that is operating abnormally. In this regard, it is also important to distinguish two classes of malfunctioning: outages and impairment. The first class comprises failures that compromise the functionality of the entire PV system, or a large part of it. For example, an inverter failure is an outage because it prevents the operation of the PV array that is connected to it. In these cases, the impact on the energy yield can be very significant, but the impact is partially mitigated by the fact that such a failure is immediately recognized and repaired. Impairments are malfunctions that do not compromise the functionality of the entire PV system and for this reason they are much harder to identify, so they may persist for a long time before being detected. For example, in a large PV plant the failure of few PV modules might be not detected by measuring the global energy production. Or, even if the impairment is detected, it may not be economically justified to replace the damaged part if the cost is higher than the value of the energy gained after the maintenance. All these factors are essential for the correct classification of failures and their related energy losses. However, PV energy losses are sensitive to the time taken to recognize the failure and to the time taken to fix it, so in order to obtain an average value of such losses, a large number of defective PV plants must be considered.

Golnas collated a large database containing information on six hundred PV plants in four continents, and data on the kinds of failure they were subject to and the corresponding energy losses [21]. The peak power rating of the monitored PV systems ranged from few kWp to 70 MWp; the oldest system was constructed in 2005. The plants contained more than 1500 inverters from 16 vendors and more than 2.2 million PV modules from 35 manufacturers.

The failure conditions were recorded via analysis of tickets registered by the PV operators or by the monitoring system. Each record contained relevant information concerning the malfunctioning occurring on the PV plant: the time of discovery, the system and device impacted, notes from the field personnel, the production impact, cost of service, and so on. There were more than 3500 tickets collected over 27 months. Although more than 50% of the monitored PV plants (350 out of 600) exhibited at least one failure, the energy production loss due to the issues captured in the tickets was 6.5 GWh, corresponding to less than 1% of the energy produced during the period under investigation.

The types of malfunction were collated into ten 'failure areas' and for each one the relative energy lost was calculated. Table 4.2 shows the most significant results.

It is worth nothing that outages involving critical subsystems encompass 69% of the identified failures and are responsible for 75% of the associated energy losses. Module failures are a very small percentage of the reported tickets and the corresponding energy loss is almost negligible. However, the low percentage of tickets activated by PV module failures is in part due to the fact that multiple PV module failures are often included in one ticket event. The analysis of tickets associated with PV failures allow different types of damage to be distinguished, as shown in Table 4.3.

The analysis of PV failures also allows construction of statistical models that are useful to establish the risks associated with PV investments, for example the likelihood of

Table 4.2 Frequency of tickets and associated energy loss for each failure area.

	Percentages	
Failure area	**of tickets**	**of kWh lost**
Inverter	43	36
AC subsystem	14	20
External*	12	20
Other	9	7
Support structure	6	3
DC subsystem	6	4
Planned outage	5	8
Modules	2	1
Weather station	2	0
Meter	1	0

Failures manifested beyond the generation meter.

Table 4.3 Frequency of tickets for specific failures of PV modules.

Failure	**Percentage of tickets**
Top glass	41
Cells	12
Backsheet	8
Connectors	6
Other	6
Junction box	6
Cables	4
Bypass diode	3

failures in domestic PV systems [22]. It has been shown that, for successful operations, it is advisable to carry out at least monthly performance checks; otherwise more than 10% of the energy may be lost because of system downtime.

References

1 SolarProfessional (2015) Module binning web pages. URL: http://solarprofessional.com/articles/design-installation/quantifying-the-impact-of-module-binning.

2 Jordan, D., Wohlgemuth, J., and Kurtz, S. (2012) Technology and climate trends in PV module degradation, in *27th European Photovoltaic and Solar Energy Conference*, Frankfurt, Germany, pp. 1–8.

3 Chamberlin, C.E., Rocheleau, M.A., Marshall, M.W., Reis, A.M., Coleman, N.T., and Lehman, P. (2011) Comparison of PV module performance before and after 11 and 20 years of field exposure, in *Photovoltaic Specialists Conference (PVSC), 2011 37th IEEE*, pp. 000 101–000 105, doi:10.1109/PVSC.2011.6185854.

4 Dolara, A., Leva, S., Manzolini, G., and Ogliari, E. (2014) Investigation on performance decay on photovoltaic modules: Snail trails and cell microcracks. *Photovoltaics, IEEE Journal of*, **4** (5), 1204–1211, doi:10.1109/JPHOTOV.2014.2330495.

5 Andrews, R.W. and Pearce, J.M. (2013) The effect of spectral albedo on amorphous silicon and crystalline silicon solar photovoltaic device performance. *Solar Energy*, **91** (0), 233–241, doi:10.1016/j.solener.2013.01.030.

6 Deline, C. (2009) Partially shaded operation of a grid-tied PV system, in *Photovoltaic Specialists Conference (PVSC), 2009 34th IEEE*, pp. 001 268–001 273, doi:10.1109/PVSC.2009.5411246.

7 Woyte, A., Richter, M., Moser, D., Mau, S., Reich, N., and Jahn, U. (2013) Monitoring of photovoltaic systems: good practices and systematic analysis, in *European PV Solar Energy Conference and Exhibition (EUPVSEC), 2013 28th*, pp. 3686–3694.

8 Tigo. Web page. URL: http://www.tigoenergy.com/it/downloads/whitepapers/.

9 Sunpower Corp. Sunpower module degradation rate. URL: http://us.sunpower.com/sites/sunpower/files/media-library/white-papers/wp-sunpower-module-degradation-rate.pdf.

10 ST Microelectronics. Calculation of reverse losses in a power diode. URL: http://www.st.com/st-web-ui/static/active/en/resource/technical/document/application_note/DM00044087.pdf.

11 Haeberlin, H. and Kaempfer, M. (2008) Measurement of damages at bypass diodes by induced voltages and currents in PV modules caused by nearby lighting currents with standard waveform, in *Photovoltaic Solar Energy Conference, 2008 23rd*, pp. 1–14.

12 IEA Photovoltaic Power Systems Programme. Review of failures of PV modules. URL: http://iea-pvps.org/index.php?id=275.

13 d'Alessandro, V., Guerriero, P., and Daliento, S. (2014) A simple bipolar transistor-based bypass approach for photovoltaic modules. *Photovoltaics, IEEE Journal of*, **4** (1), 405–413, doi:10.1109/JPHOTOV.2013.2282736.

14 Simon, M. and Meyer, E.L. (2010) Detection and analysis of hot-spot formation in solar cells. *Solar Energy Materials and Solar Cells*, **94** (2), 106–113, doi:10.1016/j.solmat.2009.09.016.

15 Geisemeyer, I., Fertig, F., Warta, W., Rein, S., and Schubert, M. (2014) Prediction of silicon PV module temperature for hot spots and worst case partial shading situations using spatially resolved lock-in thermography. *Solar Energy Materials and Solar Cells*, **120A**, 259–269, doi:10.1016/j.solmat.2013.09.016.

16 Dongaonkar, S., Deline, C., and Alam, M. (2013) Performance and reliability implications of two-dimensional shading in monolithic thin-film photovoltaic modules. *Photovoltaics, IEEE Journal of*, **3** (4), 1367–1375, doi:10.1109/JPHOTOV.2013.2270349.

17 Kim, K. and Krein, P. (2015) Reexamination of photovoltaic hot spotting to show inadequacy of the bypass diode. *Photovoltaics, IEEE Journal of*, **5** (5), 1435–1441, doi:10.1109/JPHOTOV.2015.2444091.

18 Reich, N.H., Mueller, B., Armbruster, A., van Sark, W.G.J.H.M., Kiefer, K., and Reise, C. (2012) Performance ratio revisited: is pr>90% realistic? *Progress in Photovoltaics: Research and Applications*, **20** (6), 717–726, doi:10.1002/pip.1219.

19 Jordan, D.C. and Kurtz, S.R. (2013) Photovoltaic degradation rates – an analytical review. *Progress in Photovoltaics: Research and Applications*, **21** (1), 12–29, doi:10.1002/pip.1182.

20 Leloux, J., Narvarte, L., and Trebosc, D. (2012) Review of the performance of residential PV systems in France. *Renewable and Sustainable Energy Reviews*, **16** (2), 1369–1376, doi:10.1016/j.rser.2011.10.018.

21 Golnas, A. (2012) PV system reliability: an operator's perspective, in *Photovoltaic Specialists Conference (PVSC), Volume 2, 2012 IEEE 38th*, pp. 1–6, doi:10.1109/PVSC-Vol2.2012.6656744.

22 Perdue, M. and Gottschalg, R. (2015) Energy yields of small grid connected photovoltaic system: effects of component reliability and maintenance. *IET Renewable Power Generation*, **9** (5), 432–437, doi:10.1049/iet-rpg.2014.0389.

5

Models of PV Arrays under Non-homogeneous Conditions

In the previous chapters the PV generator has been modeled by assuming that all the cells work in the same conditions and they are characterized by exactly the same values of the parameters of the SDM. This means that parametric tolerances and drifts from cell to cell are neglected, so that the SDM circuit parameters $\{I_{ph}, I_o, \eta, R_s, R_h\}$ assume exactly the same value for any cell in the array. Additionally, it has been assumed that exogenous variables – the irradiance and the temperature – also assume the same value for all the cells. As already discussed in Chapter 1, the cell behavior is simply scaled up to obtain the behavior of the whole PV field. All the current values are multiplied by the number N_p of strings connected in parallel, all the voltages are multiplied by the number N_s of cells in series connection and the two model resistances are multiplied by the factor $\frac{N_s}{N_p}$.

Unfortunately, when all these assumptions do not hold, which is in almost all real conditions, especially when PV field aging starts to have a significant effect on module performance, or when inhomogeneities of the irradiance and temperature values occur, the simple scaling operation outlined above cannot be used. This makes the PV field modeling much more complicated and its simulation quite involved and time-consuming. In principle, each PV cell might be modeled using the same SDM but with a different set of parameters $\{I_{ph}, I_o, \eta, R_s, R_h\}$, so that the level of granularity required in the simulation could be at the cell level. It is evident that, due to the very large number of cells composing even a small PV field, such a model would include a huge number of non-linear equations. Moreover, in order to describe phenomena, such as hot spots, that would be triggered by the mismatched conditions, the non-linear equations describing each cell would need to be even more complicated than those derived from the SDM.

In view of these complications, in this chapter the analysis is conducted at a module level of granularity. This means that it is assumed that the cells in each module work in homogeneous conditions: they have the same parameter values, are subject to the same irradiance levels, and work at the same temperature. The working conditions and parameters may, however, differ from module to module. In this way, for instance, a PV panel including three modules, each one protected by a bypass diode, is disassembled into three parts, each described by a non-linear equation, including the bypass diode behavior, and characterized by its own set of working parameter values. This granularity level is a good compromise between accuracy and computational burden. The proposed approach can be implemented in any simulation environment, including power-electronics simulation software such as PSIM [1], in which the user

Photovoltaic Sources Modeling, First Edition. Giovanni Petrone, Carlos Andrés Ramos-Paja and Giovanni Spagnuolo.

Companion Website: www.wiley.com/go/petrone/Photovoltaic_Sources_Modeling

can take advantage of the functional blocks or the definition of new devices through a dynamic linked library. The approach proposed in this chapter gives a clear advantage, especially in terms of simulation time, over the PV models that are already embedded in many commercial software packages and simulation environments. Moreover, commercial software requires an additional capacitance to be put in parallel with each PV module of the array in order to ensure the numerical convergence of the simulation [1].

In this chapter it is shown that, thanks to a proper symbolic manipulation of the resulting system of equations, the simulation is speed up significantly.

5.1 The use of the Lambert W-Function

The base element of the PV field representation and modeling is the module, which includes a number of series-connected cells and one bypass diode connected in anti-parallel with this string of cells. In Figure 5.1 a PV string, made up of a number of series-connected modules, is shown. The string also includes the blocking diode, which is commonly used in order to avoid current back-flows. The analysis focuses on one string only. The I–V curve of any PV field, consisting of a number of strings in parallel, results from summing the strings' current values during a voltage sweep of their I–V curves.

Each module is modeled with the SDM shown in Figure 5.2, which also includes the bypass diode D_b [2]. The non-linear system of equations (5.1)–(5.5) gives the current I

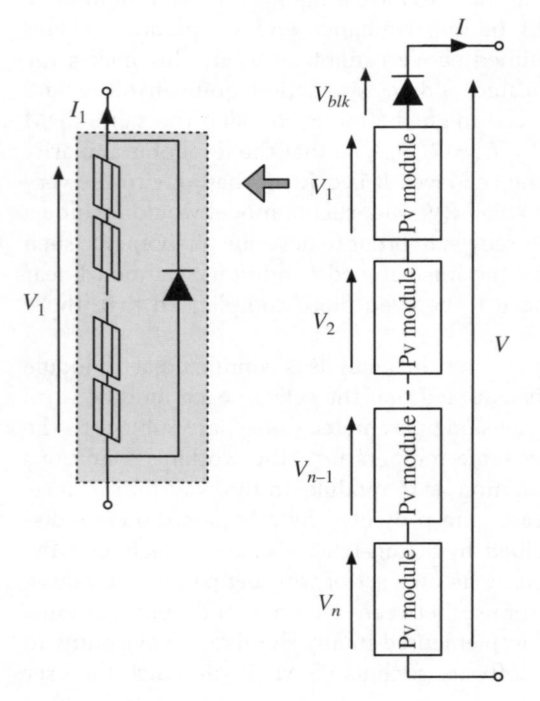

Figure 5.1 String of M series-connected PV modules including a blocking diode to prevent current back-flow.

Figure 5.2 Circuit model of a PV module including the bypass diode D_b. Source: Petrone et al. (2007) [2].

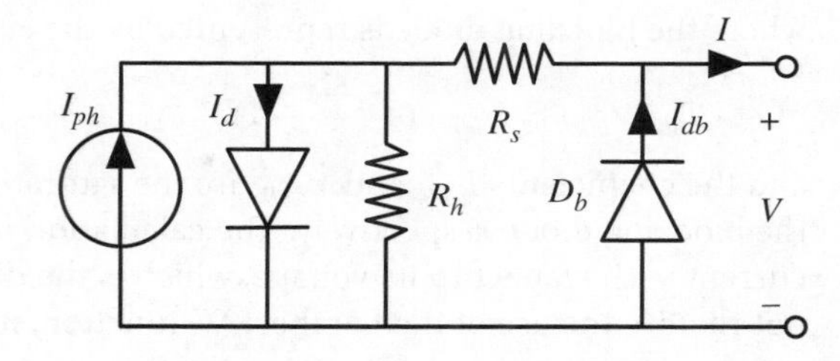

as an implicit function of the voltage V, and it includes the exogenous parameters T and G, as well as the PV module parameters.

$$I_d = I_{s,d}(e^{V_d/V_{t,d}} - 1) \tag{5.1}$$

$$I_{db} = I_{s,db}(e^{-V/V_{t,db}} - 1) \tag{5.2}$$

$$I = I_{db} + I_{ph} - I_d - I_h \tag{5.3}$$

$$V_d = V + R_s I_s = V + R_s(I - I_{db}) \tag{5.4}$$

$$I_h = \frac{V_d}{R_h} = \frac{V + R_s(I - I_{db})}{R_h} \tag{5.5}$$

Couples $\{V_{t,d}, I_{s,d}\}$ and $\{V_{t,db}, I_{s,db}\}$ are the thermal voltages and the saturation currents of the PV module and of the bypass diode, respectively. I_{ph} is the photo-induced current. Each module includes N_s series-connected cells, so that the module thermal voltage and the resistances of the whole module are obtained by multiplying the cell values by N_s.

The Lambert W-function [3], in the following represented by $W(\cdot)$, allows an explicit expression for the string current and its derivative to be obtained. These are given in (5.6) and (5.7).

$$I = \frac{R_h(I_{ph} + I_{s,d}) - V}{R_h + R_s} + I_{s,db}(e^{-V/V_{t,db}} - 1) - \frac{V_{t,d}}{R_s} W(\theta) \tag{5.6}$$

$$\frac{\partial I}{\partial V} = -\frac{1}{R_h + R_s} - \frac{I_{s,db}}{V_{t,db}}(e^{-V/V_{t,db}}) - \frac{R_h}{R_s(R_s + R_h)} W(\theta) \tag{5.7}$$

where:

$$\theta = \frac{\frac{R_s \cdot R_h}{R_s + R_h} I_{s,d} \cdot e^{\frac{R_h R_s (I_{ph} + I_{s,d}) + R_h \frac{V}{V_{t,d}} \cdot (R_h + R_s)}{}}}{V_{t,d}} \tag{5.8}$$

Equation 5.6 gives the I–V relationship for each module in the string of Figure 5.1. The whole string is modeled by the system of non-linear equations $\{F(1), F(2), \ldots, F(M), F(M+1)\}$, as shown in (5.9):

$$F(X) = \begin{cases} F(1) = V_1 + V_2 + \ldots + V_M + V_{blk} - V_{string} = 0 \\ F(2) = I_1(V_1) - I_2(V_2) = 0 \\ F(3) = I_1(V_1) - I_3(V_3) = 0 \\ \ldots \\ F(M) = I_1(V_1) - I_M(V_M) = 0 \\ F(M+1) = I_1(V_1) - I_{blk}(V_{blk}) = 0 \end{cases} \tag{5.9}$$

where the blocking diode is represented by the classical Shockley equation:

$$I_{blk}(V_{blk}) = I_{s,blk} \cdot (e^{-\frac{V_{blk}}{V_{t,blk}}} - 1) \tag{5.10}$$

and the coefficients $I_{s,blk}$ and $V_{t,blk}$ are the saturation current and the thermal voltage of the blocking diode respectively. The calculation of the derivative of the blocking diode current with respect to its voltage, which is the differential resistance, in this case does not require the use of the Lambert W-function; it is:

$$\frac{\partial I_{blk}}{\partial V_{blk}} = -\frac{I_{s,blk}}{V_{t,blk}} \cdot e^{-\frac{V_{blk}}{V_{t,blk}}} \tag{5.11}$$

The unknown variables in the system (5.9) are the module voltages and the voltage across the blocking diode V_{blk}. This means that the vector of unknown variables is $X = \{V_1, V_2, ..., V_M, V_{blk}\}$.

The solution of this system – with an assigned couple $\{G, T\}$, an assigned set of parameter values $\{I_{ph}, I_{s,d}, V_{t,d}, R_s, R_h, I_{s,db}, V_{t,db}, I_{s,blk}, V_{t,blk}\}$ for each module in the PV string, and a given value of the string voltage V_{string} – gives the voltage distribution over the modules in the string and so allows the corresponding string current for the assigned string operating conditions to be calculated. The solution of the non-linear system of (5.9) by means of the Newton–Raphson algorithm (sometimes simply referred to as "Newton's method") requires the iterative use of the relationship:

$$X_{k+1} = X_k - J^{-1}(X_k) \cdot F(X_k) \tag{5.12}$$

starting from a guess solution X_0 and with a number of iterations ensuring the convergence to a solution [4]. This means that the system of non-linear equations (5.13), having the vector of unknowns equal to $(X_{k+1} - X_k)$ must be solved repeatedly:

$$J(X_k) \cdot (X_{k+1} - X_k) = F(X_k) \tag{5.13}$$

Thus the Jacobian matrix J must be inverted many times to simulate the string, with a huge computational burden, especially in the analysis of large PV arrays. The possibility of calculating the inverse of the Jacobian matrix in a symbolic form would allow the computational problem to be simplified. The repeated solution of (5.13) at different values of X_k can be replaced with the simple matrix-vector multiplication of (5.12), in which J^{-1} is symbolically derived and the values X_k at each iteration are simply substituted in the symbolic inverse Jacobian matrix.

The way in which the topological equation in the system of (5.9) has been written has given an *arrow head* shape to its Jacobian matrix (5.14), which is also a highly sparse one:

$$J = \begin{pmatrix} 1 & 1 & 1 & \cdots & 1 & 1 & 1 \\ \frac{\partial I_1}{\partial V_1} & -\frac{\partial I_2}{\partial V_2} & 0 & \cdots & 0 & 0 & 0 \\ \frac{\partial I_1}{\partial V_1} & 0 & -\frac{\partial I_3}{\partial V_3} & 0 & 0 & 0 & 0 \\ \frac{\partial I_1}{\partial V_1} & 0 & 0 & \cdots & 0 & 0 & 0 \\ \frac{\partial I_1}{\partial V_1} & 0 & 0 & 0 & -\frac{\partial I_{M-1}}{\partial V_{M-1}} & 0 & 0 \\ \frac{\partial I_1}{\partial V_1} & 0 & 0 & 0 & 0 & -\frac{\partial I_M}{\partial V_M} & 0 \\ \frac{\partial I_1}{\partial V_1} & 0 & 0 & 0 & 0 & 0 & -\frac{\partial I_{blk}}{\partial V_{blk}} \end{pmatrix} \tag{5.14}$$

It is worth noting that all the non-zero elements in (5.14) are numerically calculated, using (5.7) and (5.11), at the given module operating conditions and for the assigned voltage set X_k of the current Newton–Raphson iteration (Equation 5.12).

The concept of differential resistance is useful for a compact formalization of the model, by including the modules and the blocking diode:

$$\frac{\partial I_1}{\partial V_1} = \frac{1}{R_1}, \ldots, \frac{\partial I_n}{\partial V_n} = \frac{1}{R_n} \tag{5.15}$$

and

$$\frac{\partial I_{blk}}{\partial V_{blk}} = \frac{1}{R_{blk}} \tag{5.16}$$

By using these definitions, the Jacobian matrix of (5.14) can be partitioned into the sub-matrices shown below. In particular:

$$\mathbf{A} = [1]_{1\times 1} \tag{5.17}$$

is the integer "1",

$$\mathbf{B} = [1\,1\,1\ldots 1\,1\,1]_{1\times M} \tag{5.18}$$

is a row vector of length M, with all the elements equal to 1.

$$\mathbf{C} = \frac{1}{R_1}[1\,1\,1\ldots 1\,1\,1]^T_{1\times M} \tag{5.19}$$

is a column vector of length M, with all the elements equal to $\frac{1}{R_1}$, and

$$\mathbf{D} = \begin{pmatrix} -\frac{1}{R_2} & 0 & 0 & 0 & 0 & 0 \\ 0 & -\frac{1}{R_3} & 0 & 0 & 0 & 0 \\ 0 & 0 & \cdots & 0 & 0 & 0 \\ 0 & 0 & 0 & \cdots & 0 & 0 \\ 0 & 0 & 0 & 0 & -\frac{1}{R_n} & 0 \\ 0 & 0 & 0 & 0 & 0 & -\frac{1}{R_{blk}} \end{pmatrix}_{M\times M} \tag{5.20}$$

is a diagonal square matrix of size n. By using these blocks, the Jacobian matrix of (5.14) is expressed as:

$$\mathbf{J} = \begin{pmatrix} \mathbf{A} & \mathbf{B} \\ \mathbf{C} & \mathbf{D} \end{pmatrix} \tag{5.21}$$

The block representation in (5.21) allows the inverse of the Jacobian matrix, which is mandatory for solving (5.13), to be calculated symbolically by means of the Schür complement tool [5]. The Schür complement S of J is defined as:

$$\mathbf{S} \stackrel{def}{=} \mathbf{A} - \mathbf{B}\mathbf{D}^{-1}\mathbf{C} \tag{5.22}$$

and in the case under study it is a scalar. Using this definition results in the inverse matrix of the Jacobian J assuming the following form [5]:

$$\mathbf{J}^{-1} = \begin{pmatrix} \mathbf{S}^{-1} & -\mathbf{S}^{-1}\mathbf{B}\mathbf{D}^{-1} \\ -\mathbf{D}^{-1}\mathbf{C}\mathbf{S}^{-1} & \mathbf{D}^{-1} + \mathbf{D}^{-1}\mathbf{C}\mathbf{S}^{-1}\mathbf{B}\mathbf{D}^{-1} \end{pmatrix} \tag{5.23}$$

The matrix expression in (5.23) requires simple matrix operations; also the evaluation of the inverse of the matrix D is trivial, because it is diagonal (see (5.20)). Using (5.17)–(5.20), the Schür complement S is a real number:

$$S = 1 + \frac{1}{R_1} \cdot (R_2 + R_3 + \ldots + R_M + R_{blk}) \tag{5.24}$$

Finally, by defining the differential resistance:

$$R_{equiv} = \sum_{i=1}^{n} R_i + R_{blk} \tag{5.25}$$

the inverse of the Jacobian matrix becomes:

$$\mathbf{J}^{-1} = \frac{1}{R_{equiv}} \mathbf{M}_R \tag{5.26}$$

where

$$\mathbf{M}_R = \begin{pmatrix} R_1 & R_1R_2 & R_1R_3 & . & . & R_1R_{blk} \\ R_2 & R_2^2 - R_2R_{equiv} & R_2R_3 & . & . & R_2R_{blk} \\ R_3 & R_3R_2 & R_3^2 - R_3R_{equiv} & R_3R_4 & . & R_3R_{blk} \\ . & . & . & . & . & . \\ . & . & . & . & . & . \\ R_{blk} & R_{blk}R_2 & R_{blk}R_3 & . & . & R_{blk}^2 - R_{blk}R_{equiv} \end{pmatrix}_{(M+1)\times(M+1)}$$

In conclusion, once the operating conditions, the values of the SDM parameters of the modules in Figure 5.1, and the string voltage V_{string} have been assigned, the string current value is calculated by applying the iterative Newton–Raphson formula of (5.12). This computation is sped up by using the symbolic expression of the matrix $J^{-1}(X_k)$ obtained in (5.26), which is calculated by substituting the PV string parameters and operating condition values in (5.15) and (5.16), meaning (5.7) and (5.11) for the current X_k value.

The next section shows how this model can be used for simulating a PV string. The calculation of the whole I–V curve of a given string of an assigned set of PV modules subject to a number of different irradiance levels is illustrated. Moreover, the calculation of a single value of the I–V curve, namely the evaluation of the string current at one assigned value of its voltage, is also outlined. These two problems seem to be the same but, as is explained in the following sections, this is not completely true.

5.2 Application Examples

5.2.1 The Entire I–V Curve of a Mismatched PV String

As a first example, the calculation of the whole I–V curve of a PV string working in mismatched conditions is outlined. A sequence of increasing V_{string} values is applied in a range between 0 V and the open-circuit voltage of the string. The performance of the proposed approach, which applies the iteration of (5.12) using the explicit expression (5.26) of the inverse of the Jacobian matrix, is compared with that derived from a reference method for solving the system of non-linear equations (5.9), namely the MATLAB® *fsolve* function [6]. It is well known that *fsolve* takes as its input the system of equations and the initial guess solution. The Jacobian matrix can be given as a further input in order to save some computation time. The time needed to reach the solution is reported in both cases, because the expression of the Jacobian matrix (5.14) is available.

The choice of the initial guess solution, which is required by the Newton–Raphson iteration (5.12) as well as by the *fsolve* MATLAB® algorithm, deserves some comment because it affects the convergence of any iterative algorithm [4]. Commercially available high-FF cells typically have a very high shunt resistance R_h, and thus a very small $\partial I/\partial V$ at very low voltage values. As a consequence, the problem may be not well-conditioned. Moreover, at low V_{string} values, the guess solution is not immediately determined, because only the modules at the highest short-circuit current value in the studied string work, and the others are bypassed. Instead, the I–V curve calculation can be started from the string open-circuit conditions, where all the modules work at their own open-circuit voltage V_{oc}. The guess solution is determined by means of (5.27), which is a good approximation of each module's open-circuit voltage for an infinite R_h value [7]:

$$V_{oc} = V_{t,d} \cdot \ln(1 + \frac{I_{ph}}{I_{s,d}}) \tag{5.27}$$

With this guess solution, the iteration (5.12) is performed, until convergence is reached at the string open-circuit voltage. Then, as in a chain, the guess solution for the next point, at the chosen lower voltage value, is fixed as the final solution obtained at the end of the iteration performed for the preceding voltage value, as just calculated.

The performance of the proposed algorithm – in the following denoted NRiJ, standing for Newton–Raphson with an explicit inverse of the Jacobian matrix – is now compared with those obtained from MATLAB® through the *fsolve* function. The *fsolve* algorithm, which is denoted FS in the following, is considered a reference because it can be used by simply defining the system of equations and without any symbolic calculation effort.

The same *fsolve* MATLAB® algorithm has been also used by giving to it as a further input the explicit Jacobian matrix. This allows the computational advantage in providing the symbolic expression of the Jacobian matrix (5.14) to be shown, especially in comparison with the proposed method, which profits from the availability of the inverse of the Jacobian matrix. This approach is referred to as TRDJ in the following, because it uses *fsolve* MATLAB® with the trust-region dog-leg method with an explicit Jacobian matrix.

In order to do a fair comparison, the termination condition for these three methods is fixed with the same value of tolerance, equal to 10^{-6}. Simulation results are obtained for modules of $N_s = 32$ cells each, having a photo-induced current calculated through the translational formula described in Chapter 3 (3.4), with $G_0 = 1\,\mathrm{kW/m^2}$, $I_{ph,0} = 6.9\,\mathrm{A}$, $\alpha_i = 0.0017\,\mathrm{A/K}$ and $T_0 = 298\mathrm{K}$. The module saturation current is determined through the translational formula (3.15), with $T_0 = 298\mathrm{K}$, $I_{s0} = 3.0375 \times 10^{-9}\,\mathrm{A}$ and the energy gap value considered as independent from temperature, so that $E_g/k = 1.2839 \times 10^4\mathrm{K}$. The modules' series and shunt resistances have been fixed at $R_s = 6\,\mathrm{m\Omega}$ and $R_h = 10\,\mathrm{k\Omega}$, respectively. The PV diode ideality factor is fixed at $\eta = 1.35$. The values used for the bypass and blocking-diode saturation current and ideality factors are $I_{s,db} = I_{s,blk} = 1 \times 10^{-14}\,\mathrm{A}$ and $\eta_{db} = \eta_{blk} = 1.2$ respectively. All these values are used in (5.6) for the computation of the current flowing in each module.

Table 5.1 Irradiation and temperature profile for the simulated PV string.

Module number	Irradiation [W/m²]	Temperature [K]
1	1000	345
2–3	700	320
4	100	298
5–7	250	300
8–9	850	315
10–11	530	306
12–13	620	310
14–16	715	312
17–18	400	305
19–20	260	302
21–22	575	312
23–24	500	307
25–27	950	322
28–29	755	316
30	280	301

The performances of the algorithms are evaluated on an example concerning a string of 30 modules working at 15 different irradiation levels and temperatures (see Table 5.1). Figure 5.3 shows the string I–V curve reconstructed in 1516 voltage samples using the three methods. The curves obtained by the three methods are so close that the differences are indistinguishable. Using a AMD Phenom 9650 quad-core 2.31-Ghz processor with 1.93 GB of RAM, the NRiJ algorithm reaches the right solution in about 12 s, which is less than the 47% of the time needed by the TRDJ method, and with a comparable mean-square percentage error.

In terms of the number of calls to the vector function (5.9), *fsolve* requires a high number of calls at those voltage values corresponding to the inflection points, where the bypass diodes enter into the conduction mode and the slope of the I–V curve changes abruptly. The NRiJ algorithm needs 89.6% of the number of function evaluations that the TRDJ method does (2739 against 3057), but with less than 47% of the computation time needed.

5.2.2 The Operating Point of a Mismatched PV String

The straightforward calculation of the voltage distribution over the modules in a PV string $X = \{V_1, V_2, ..., V_M, V_{blk}\}$ for an assigned value of the string voltage V_{string} is a more challenging task with respect to the reconstruction of the whole I–V curve. Indeed, as explained in the previous section, the latter started from the open-circuit voltage of the string, for which the initial guess solution is determined quite easily and accurately. The intrinsic weakness of the Newton–Raphson method with respect to the guess solution makes the problem of calculating the string current in a given V_{string} value much

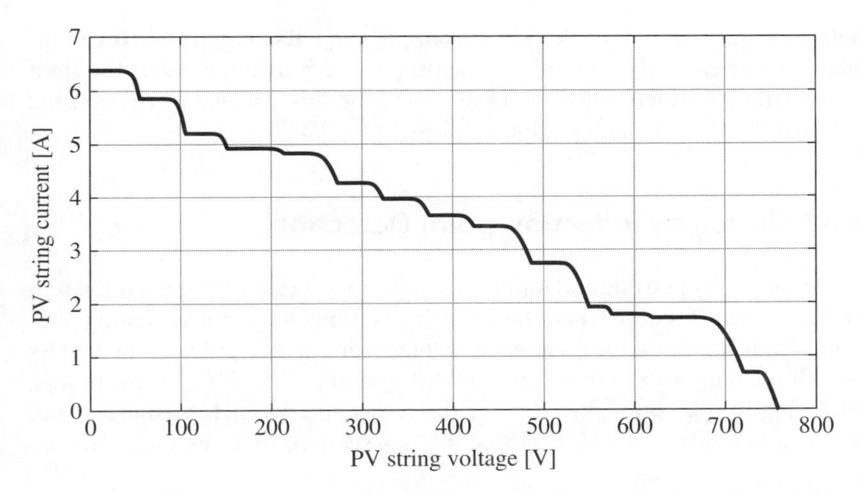

Figure 5.3 I–V curve of the PV string obtained by using the three methods. Differences are indistinguishable.

harder. Calculation of a guess solution that is quite close to the exact one and the adoption of the damped Newton algorithm [4], which employs a variable step-size (5.28), are helpful approaches to solving this problem.

The calculation of an improved guess solution is considered in detail in the next section. The proposed method is computationally inexpensive because it uses a simplified model of the PV string that is aimed at calculating a good approximation of the so-called *inflection points* of the PV string. These are the voltages at which one or more bypass diodes enters into conduction because the string current is bigger than the short-circuit current of the module that the bypass diode is anti-parallel with. The two inflection points bounding the V_{string} value at which the string current must be calculated allow a good guess solution to be determined, allowing the damped Newton method to converge safely.

The damped Newton method uses the following iterative formula:

$$X_{k+1} = X_k - \alpha_k \cdot J^{-1}F(X_k) \tag{5.28}$$

Because of the fact that the vector of unknown variables X (5.9) has length $(M+1)$, then the step-size α_k is the diagonal matrix:

$$\alpha_k = diag\left\{min\left\{1, \frac{\gamma(1)}{|\Delta X_{k+1}(1)|}\right\}, min\left\{1, \frac{\gamma(2)}{|\Delta X_{k+1}(2)|}\right\}, \ldots, \right.$$
$$\left. min\left\{1, \frac{\gamma(M+1)}{|\Delta X_{k+1}(M+1)|}\right\}\right\} \tag{5.29}$$

where $\Delta X_{k+1}(i)$, is the ith element of the vector:

$$\Delta X_{k+1} = J(X_k)^{-1} \cdot F(X_k) \tag{5.30}$$

The user-defined constant $\gamma(i)$ has the role of limiting the changes of X in the search process. Its value is fixed at $\gamma(i) = 5$ for $i = 1 \ldots M$ and $\gamma(M+1) = 0.5$: these values, obtained at the end of the experimental activity, ensure fast convergence.

The procedure is again tested on the same example used above, with 1516 voltage values calculated independently each other. The proposed algorithm reaches the right solution in less than 63 s, which is less than the 57% of the time needed by *fsolve* using TRDJ, with almost the same accuracy, which is close to 5×10^{-15}.

5.3 Guess Solution by Inflection-point Detection

The approach proposed by Petrone and Ramos-Paja [8] aims at calculating the inflection points occurring in the I–V curve when the PV string is subjected to mismatching phenomena. The inflection point appears at each voltage value where at least one bypass diode in the string changes its status. The method uses the ideal PV module model, thus without taking into account the effect of the series and parallel resistances, and considers the bypass diode as an ideal device. In this section, these assumptions are removed.

The example shown in Figure 5.4 is helpful in showing that, at the inflection point, if the short-circuit current of the jth module is greater than the short-circuit current of the kth one, namely $I_{sc,j} \geq I_{sc,k}$ with $j, k \in [1...n]$, then the kth bypass diode turns ON in the whole voltage range where the following inequality holds:

$$I_j > I_k \tag{5.31}$$

By assuming an ideal behavior of the bypass diode, the kth module is short-circuited in the whole voltage range where (5.31) holds. As a consequence, in the same voltage range, the I–V curve of the series connection of the jth and kth modules coincides with the I–V curve of the jth one. At some voltage value V_{0jk}, the condition of (5.31) does not hold anymore and the kth bypass diode turns OFF, so that for $V > V_{0jk}$ the I–V curve of the series connection is a combination of the I–V curves of both the jth and the kth modules. The voltage V_{0jk} at which the inflection point occurs is calculated by taking into account *all* the following conditions:

$$I_j = I_k \tag{5.32}$$

$$V_j = V_{0jk} \tag{5.33}$$

$$V_k = 0 \tag{5.34}$$

Under the assumption that the series and parallel resistances appearing in the SDM are $R_s = 0$ and $R_h \to \infty$ and that the bypass diodes are ideal, the ideal PV module I–V relationship for the two modules are:

$$I_j = I_{phj} - I_{s,dj}(e^{V_j/V_{t,dj}} - 1) \tag{5.35}$$

$$I_k = I_{phk} - I_{s,dk}(e^{V_k/V_{t,dk}} - 1) \tag{5.36}$$

By taking into account all the conditions (5.34)–(5.36), the voltage at which the inflection point occurs is calculated as follows:

$$V_{0jk} = V_{t,dj} \cdot \ln(\frac{I_{phj} + I_{s,dj} - I_{phk}}{I_{s,dj}}) \tag{5.37}$$

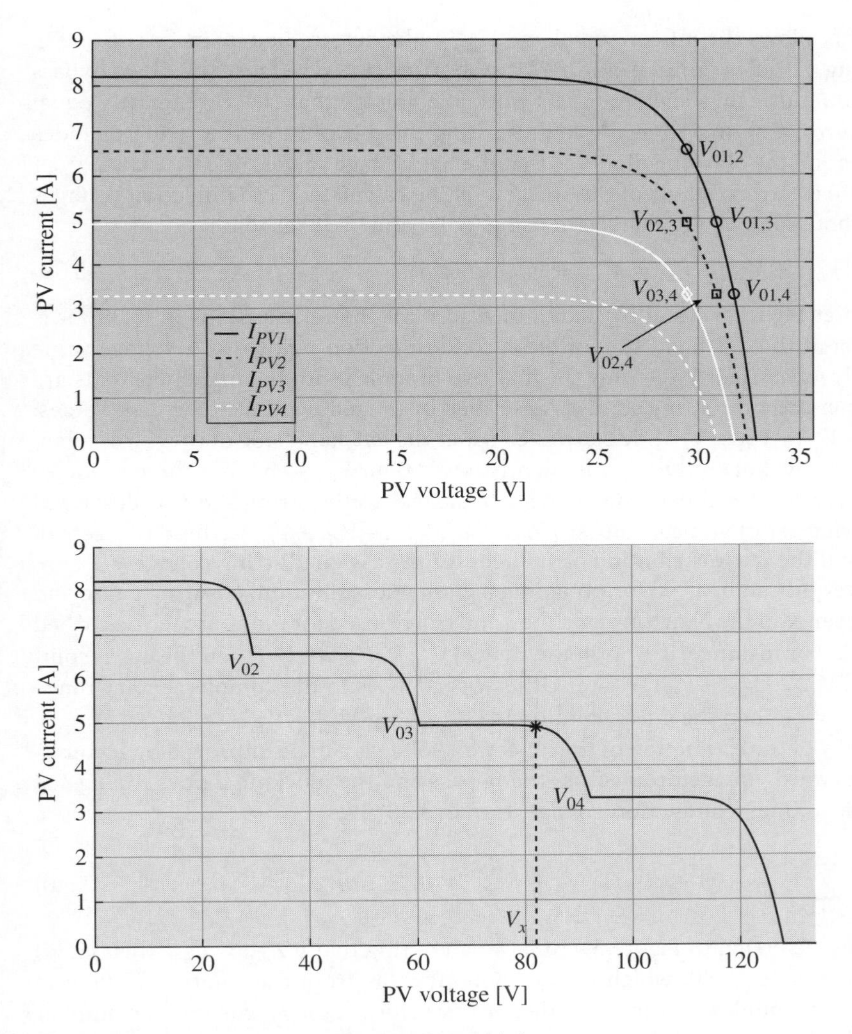

Figure 5.4 Inflection points for four PV series modules.

As $R_h \to \infty$, $I_{sc,k} \approx I_{ph,k}$, and the PV modules in the string are first sorted so that $I_{sc,k-1} > I_{sc,k}$ and then, the kth module becomes active at the voltage value:

$$V_{0k} = \sum_{m=1}^{k-1} V_{0mk} \qquad k \in [2, .., M] \tag{5.38}$$

where the terms in the sum are calculated according to (5.37). V_{0k} is the *inflection voltage*, which is the minimum string voltage at which the kth module becomes active. V_{0k} is used for determining an improved guess solution for Newton's method.

Equation 5.38 allows every inflection voltage occurring in the range $[0, V_{oc}]$ of a mismatched PV string to be calculated. So, given a voltage value $V_x \in [0, V_{oc}]$, it is possible to determine the two inflection voltages bounding V_x. Indeed, the first m modules are active and this allows the inflection voltage V_{0mk} to be determined by means of (5.37). The remaining modules, from $m + 1$ to M, are short-circuited because their own bypass diode is conducting. Thus such modules work at a voltage that is approximately equal to $-V_{db}$. Figure 5.4 shows an example illustrating this procedure. The string includes four modules and the blocking diode, so that the five voltage values $\{V_1, V_2, V_3, V_4, V_{blk}\}$ describing the electrical status of the string must be calculated. The inflection voltages V_{03} and V_{04} bounding V_x are calculated using (5.37) and (5.38) as follows:

$$V_{03} = V_{013} + V_{023} \quad V_{04} = V_{014} + V_{024} + V_{034} \tag{5.39}$$

By looking at Figure 5.4 and by taking into account the relationships in (5.39), it is clear that just at the left-hand side of the second inflection point – at a voltage value that is slightly lower than V_{03} – only the first two modules with the higher currents are producing energy, so the string status is described by the following set of voltage values: $\{V_{013}, V_{023}, -V_{db}, -V_{db}, -V_{blk}\}$. Meanwhile, just at the left-hand side of the third inflection point – at a voltage value that is slightly lower than V_{04} – the first three modules produce energy and the fourth one alone is bypassed, so the string status is described by the following set of voltage values: $\{V_{014}, V_{024}, V_{034}, -V_{db}, -V_{blk}\}$. These two sets of voltages bound the real distribution of voltages for the assigned string voltage V_x.

Such a criterion can be used for obtaining a good guess solution, ensuring a safe and quick convergence of the Newton algorithm, and can be used for the approach described in Section 5.2. For instance, if V_x is in the range $[V_{03}, V_{03} + \frac{V_{04}-V_{03}}{2}]$ then the guess solution is $\{V_{013}, V_{023}, -V_{db}, -V_{db}, -V_{blk}\}$. Otherwise, if V_x is in the complementary range $[V_{03} + \frac{V_{04}-V_{03}}{2}, V_{04}]$, then the guess solution is $\{V_{014}, V_{024}, V_{034}, -V_{db}, -V_{blk}\}$.

The accuracy of the estimation of the inflection voltages can be improved by accounting for the forward voltage drop across the bypass and the blocking diodes. Using V_{db} and V_{blk} as the diodes' conduction voltage drop in 5.38 gives:

$$V_{0k} = \sum_{m=1}^{k-1} V_{0mk} - (n - k + 1) \cdot V_{db} - V_{blk}\ , k \in [2, .., n] \tag{5.40}$$

For example, referring to Figure 5.4, it is worth noting that if $I_{ph,2} = I_{ph,3}$, then (5.37) gives the value of $V_{023} = 0$, which is poor and far away from the solution. Taking in account that the modules are sorted for decreasing values of the I_{ph} current, the number of modules having the same I_{ph} current as the inflection point just preceding the V_x voltage must be determined. Therefore, by assuming that V_{0m} is the inflection point immediately lower than the V_x voltage, then the number of modules q that have the same I_{ph} current as module m is calculated. The inflection point just greater than the V_x voltage will be $V_{0(m+q)}$, so the guess solution for the modules m to $(m + q - 1)$ must be between the voltages evaluated in (5.37), which generates V_{0m}, and those evaluated in (5.37), which generates the inflection point $V_{0(m+q)}$.

5.4 Real-time Simulation of Mismatched PV Arrays

Real-time simulation of a mismatched array is useful in many contexts. The one that is most frequently discussed in literature is related to the emulation – which is not simulation – of a PV generator using a switching DC/DC converter. An embedded device is used to run a model of the PV generator and so to control a power converter that emulates the I–V relationship that is typical of the PV source. The converter is therefore used as a sort of amplifier of the signal produced by the embedded device. A limit of this application is in the fact that the steady-state behavior of the PV source is accurately emulated [9] but the dynamic behavior, emulating an irradiance variation, is reproduced in a way that differs from the real behavior. This is mainly due to the DC/DC converter dynamics.

A different application is shown in a limited number of scientific contributions [10, 11]: the PV source is accurately modeled in an embedded system for the purpose of a rapid prototyping development of circuits oriented to, for example, control and real-time simulation. In order to achieve this result, various methods can be used, including those described in this chapter and especially in Section 5.3. Mellit et al. [10] used a neural network for modeling a complete standalone system, including the PV generator, the storage system, and the related switching converters. The simulation was performed on a FPGA device and partial shadowing was not simulated at all. A similar approach was adopted by Tornez-Xavier et al. [11], who discussed the modeling of the PV source only, again in uniform operating conditions.

5.5 Estimation of the Energy Production of Mismatched PV Arrays

The models presented in Chapters 4, 5, and 6 have been used in the literature to estimate the energy production of mismatched PV arrays to achieve different balances between complexity, accuracy, and calculation speed.

The simplest mathematical/physical model of mismatched PV arrays used to perform energy-yield evaluations uses the ideal SDM (ISDM), discussed in Section 2.3, to represent each module forming the array [8]. Since the ISDM has an explicit relation between the PV current and voltage, the approach proposed in the paper is based on decomposing the Jacobian formed by the set of non-linear equations adopted to model the strings, so that an explicit expression for the analytical inverse-Jacobian is obtained. That characteristic makes the model very fast to compute when using, for example, the Newton–Raphson method. However, since the model disregards the effects of both the series and parallel resistances of the SDM, it introduces prediction errors that would be avoided by use of a model based on the complete SDM, for example the model presented in this chapter. The simplified model [8] exhibits higher errors (around 5%) when compared with the model presented in this chapter, although the simplified model needs only 0.18% of the processing time required by the more complex one. Therefore, the simplified model is useful for performing, in a reasonable time, annual or lifetime energy estimates to support viability analysis. This could take months of calculations if the more complex model described in this chapter were used. Finally, the same paper [8] illustrates the model's application to energy predictions by considering a PV array made up of strings in parallel.

An improvement to this energy estimation model has been developed [12] by including two types of phenomena:

- the ISDM parameter values are recalculated each time in terms of the ambient temperature

- the model includes information concerning the shading geometry and its movement throughout the day over the PV array, so that more precise estimates can be made.

It allows an evaluation to be made of the effect of predictable shading profiles, for example those generated by buildings, trees, posts, and so on. The paper's results show an improvement of the energy prediction for the PV array over a classical, average shading pattern.

The ISDM has also been used to estimate the energy production of mismatched total cross-tied (TCT) arrays [13], using the same approach as discussed at the beginning of this subsection [8], but applied to the non-linear equation system modeling TCT electrical relations. This TCT modeling approach has also been improved by introducing temperature effects and dynamic shading profiles [12].

An alternative to increasing the precision of the energy production estimate is to adopt the complete SDM in the electrical equations instead of the ISDM. However, mismatched PV arrays must also include in the electrical equations the behavior of the bypass diodes. The most accurate approach is to model these bypass diodes using the Shockley equation, as described in Section 5.1, which introduces another non-linear component to the equation system. To reduce the calculation time, Bastidas et al. [14] adopt the SDM for each module forming the array but use a bi-linear model for the bypass diode: open circuit when the bypass diode is inactive and a resistive behavior when the bypass diode is active. With that simplification, their model is able to achieve an energy prediction with an error lower than 1% but only requiring 39.8% of the time used to compute the model described in Section 5.1.

Another approach based on the SDM introduces some simplifications to reduce the time required for calculating the current in mismatched PV strings [15]:

- all the modules have approximately the same open-circuit voltage
- the MPP currents of the mismatched string are approximately equal to the MPP currents of each module.

The MPPs of the string are then interpolated using the open-circuit voltage and MPP currents. This approach requires very short computation times in comparison with the SDM: it requires only 0.016% of the processing time required by the model presented in this chapter, with errors close to 4%. Unfortunately, this technique is applicable to single strings, and hence it is not suitable for predicting the energy production of series–parallel (SP) PV arrays, so it can be used for evaluating the viability of small (typically single-string) PV systems, or for estimating the energy production of micro-inverter-based PV installations.

The main disadvantage of these approaches concerns the errors that can be introduced by changes in the module temperature. This problem can be addressed by updating the parameter values of the SDM using experimental measurements [16]. This method has been used to predict next-day energy production [16], but only for uniformly irradiated modules. Moreover, predictions of the shade patterns, which dynamically change the mismatch conditions of the PV array [12], must be included in the more precise models to improve the prediction accuracy.

In models for mismatched operations, the temperature dependence is required to apply the tests described in the IEC 61853-1 standard: the power rating of PV modules must be performed at seven irradiance and four temperature matrix levels to accurately predict energy production of PV modules at different climatic conditions [17]. Therefore, an extension of the standard tests to mismatched PV arrays must include temperature-dependent parameters in the SDM representing each module.

The adjustment of the SDM parameters is commonly performed using the short-circuit current, open-circuit voltage, and MPP current and voltage from a single I–V curve (typically at STC), as described in Chapters 2 and 3. On the other hand, records of I–V curves at the wide range of irradiance and temperature conditions specified in the IEC 61853-1 have been used to improve the SDM parameters too [18, 19]. However, these methods consider uniformly irradiated modules, so an extension that considers mismatched PV arrays is needed.

References

1 PowerSim (2014) PSIM power conversion and control simulator. URL: http://powersimtech.com/products/psim/.

2 Petrone, G., Spagnuolo, G., and Vitelli, M. (2007) Analytical model of mismatched photovoltaic fields by means of Lambert W-function. *Solar Energy Materials and Solar Cells*, **91** (18), 1652–1657.

3 Corless, R.M., Gonnet, G.H., Hare, D.E.G., Jeffrey, D.J., and Knuth, D.E. (1996) On the Lambert W function, in *Advances in Computational Mathematics*, **5** (1), 329–359.

4 Press, W.H., Teukolsky, S.A., Vetterling, W.T., and Flannery, B.P. (2007) *Numerical Recipes: The Art of Scientific Computing*, 3rd edn. Cambridge University Press.

5 Zhang, F. (2010) *The Schur Complement and its Applications*, Numerical Methods and Algorithms, Springer US.

6 Mathworks (2014) Fsolve documentation. URL: http://www.mathworks.com/help/toolbox/optim/ug/fsolve.html.

7 Villalva, M., Gazoli, J., and Filho, E. (2009) Comprehensive approach to modeling and simulation of photovoltaic arrays. *Power Electronics, IEEE Transactions on*, **24** (5), 1198–1208.

8 Petrone, G. and Ramos-Paja, C. (2011) Modeling of photovoltaic fields in mismatched conditions for energy yield evaluations. *Electric Power Systems Research*, **81** (4), 1003–1013.

9 Koutroulis, E., Kalaitzakis, K., and Tzitzilonis, V. (2006) Development of an FPGA-based system for real-time simulation of photovoltaic modules, in *Rapid System Prototyping, 2006. Seventeenth IEEE International Workshop on*, pp. 200–208, doi:10.1109/RSP.2006.14.

10 Mellit, A., Mekki, H., Messai, A., and Salhi, H. (2010) FPGA-based implementation of an intelligent simulator for stand-alone photovoltaic system. *Expert Systems with Applications*, **37** (8), 6036–6051, doi:10.1016/j.eswa.2010.02.123.

11 Tornez-Xavier, G., Gomez-Castaneda, F., Moreno-Cadenas, J., and Flores-Nava, L. (2013) FPGA development and implementation of a solar panel emulator, in *Electrical Engineering, Computing Science and Automatic Control (CCE), 2013 10th International Conference on*, pp. 467–472, doi:10.1109/ICEEE.2013.6676052.

12 Ramos-Paja, C., Saavedra-Montes, A., and Trejos, L. (2015) Estimating the produced power by photovoltaic installations in shaded environments. *DYNA*, **82** (192), 37–43.
13 Ramos-Paja, C., Bastidas, J., Saavedra-Montes, A., Guinjoan-Gispert, F., and Goez, M. (2012) Mathematical model of total cross-tied photovoltaic arrays in mismatching conditions, in *2012 IEEE 4th Colombian Workshop on Circuits and Systems, CWCAS 2012 – Conference Proceedings.*
14 Bastidas, J., Franco, E., Petrone, G., Ramos-Paja, C., and Spagnuolo, G. (2013) A model of photovoltaic fields in mismatching conditions featuring an improved calculation speed. *Electric Power Systems Research*, **96** (0), 81–90.
15 Orozco-Gutierrez, M., Petrone, G., Ramirez-Scarpetta, J., Spagnuolo, G., and Ramos-Paja, C. (2015) A method for the fast estimation of the maximum power points in mismatched PV strings. *Electric Power Systems Research*, **121**, 115–125.
16 Accetta, G., Piroddi, L., and Ferrarini, L. (2012) Energy production estimation of a photovoltaic system with temperature-dependent coefficients, in *Sustainable Energy Technologies (ICSET), 2012 IEEE Third International Conference on*, pp. 189–195.
17 Paghasian, K. and TamizhMani, G. (2011) Photovoltaic module power rating per IEC 61853-1: a study under natural sunlight, in *Photovoltaic Specialists Conference (PVSC), 2011 37th IEEE*, pp. 002 322–002 327.
18 Hansen, C.W. (2013) Estimation of parameters for single diode models using measured IV curves, in *Photovoltaic Specialists Conference (PVSC), 2013 IEEE 39th*, pp. 0223–0228.
19 Dobos, A.P. and MacAlpine, S.M. (2014) Procedure for applying IEC-61853 test data to a single diode model, in *Photovoltaic Specialist Conference (PVSC), 2014 IEEE 40th*, pp. 2846–2849.

6

PV Array Modeling at Cell Level under Non-homogeneous Conditions

As discussed in Chapter 5, under mismatched conditions, and especially whenever deep shadowing occurs, cells in a string might be driven to work at a negative value of the voltage, with respect to the sign convention used for the cell current and voltage references. In such conditions the cell with a negative voltage and a positive current sinks electrical power, so that it, or even a part of its area, can be subject to a temperature increase and permanent damage. If the negative voltage is limited to small values, the high value of the cell shunt resistance keeps the current at low values. However, if the cells biased at a negative voltage are forced to work at a high forward current, the cells enter into breakdown. The voltage becomes a high negative value, the current is high and positive, and so the high power absorbed by the cell damages it.

The modeling of these phenomena is described in the following sections. A different model is required because, unfortunately, those described in Chapter 5 do not allow simulation of reverse behavior of cells.

6.1 PV Cell Modeling at Negative Voltage Values

A few different models have been proposed for describing cell operations in the second quadrant of the Cartesian plane in which its I-V curve is plotted [1, 4]. The SDM or the DDM discussed in the previous chapters is modified by adding a term accounting for the cell behavior in breakdown conditions, namely at high negative voltages.

6.1.1 The Bishop Term

The additional term introduced by Bishop [1] reproduces the high current in the cell during avalanche breakdown at high negative voltages. The branch of the cell equivalent circuit containing the shunt resistance is completed with a voltage-controlled current generator, as shown in Figure 6.1.

In this way, the current through the shunting branch is expressed as:

$$I = I_{ph} - I_s \cdot \left(e^{\frac{V+I\cdot R_s}{\eta V_t}} - 1\right) - \frac{V + I \cdot R_s}{R_h} \cdot \left[1 + \alpha \cdot \left(1 - \frac{V + I \cdot R_s}{V_{br}}\right)^{-m}\right] \tag{6.1}$$

where V_{br} is the junction breakdown voltage, α is the fraction of the ohmic current involved in avalanche breakdown, and m is the avalanche breakdown exponent.

Figure 6.2 gives an example of the behavior of a PV cell in both the first and the second quadrant.

Photovoltaic Sources Modeling, First Edition. Giovanni Petrone, Carlos Andrés Ramos-Paja and Giovanni Spagnuolo.

Companion Website: www.wiley.com/go/petrone/Photovoltaic_Sources_Modeling

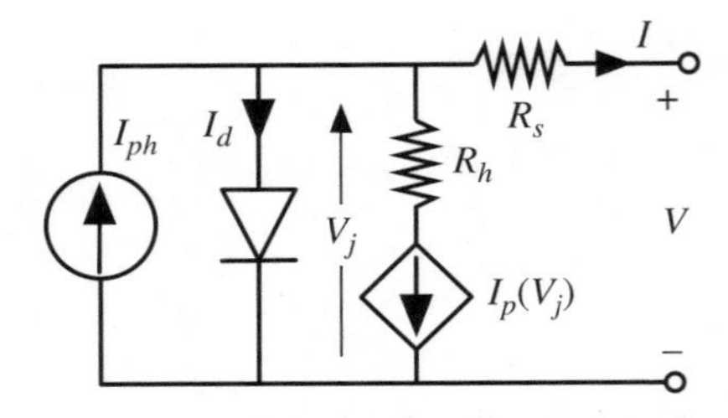

Figure 6.1 Single-diode model including the reverse polarization term.

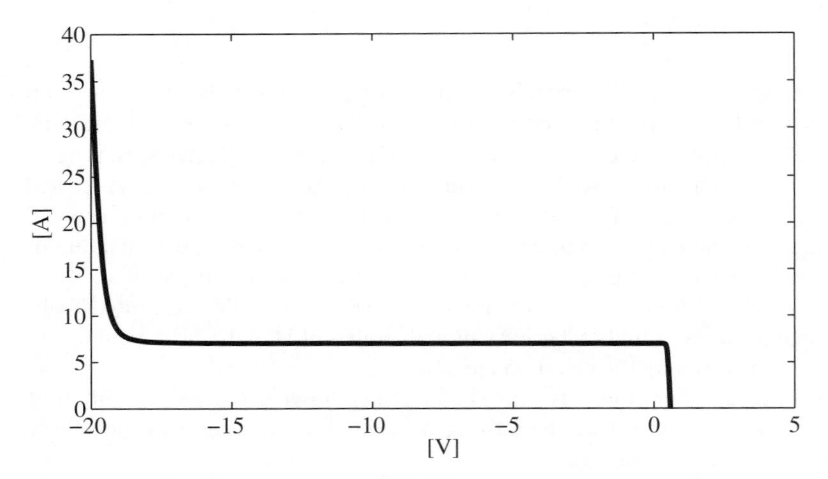

Figure 6.2 Complete I–V curve of a PV cell. This plot has been obtained by using the following parameter values: $\alpha = 0.002$, $m = 3$, $V_{br} = -20$ V, $R_s = 0.01\ \Omega$, $R_{sh} = 200\ \Omega$, $I_s = 0.135$ nA, $I_{ph} = 7$ A, $T = 273$ K.

The shunt resistance R_h has an effect on the slope of the I–V curve in the second quadrant; the higher the R_h value the flatter the curve in a voltage range from 0 V almost up to the breakdown voltage V_{br}. A large R_h value is useful when the cell works in the first quadrant, thus when it produces electrical energy, because it contributes to reaching a high FF. In other words, a high R_h value allows the I_{mpp} and I_{sc} values to be very close. On the other hand, when the cell works in reverse mode, a low R_h value would allow the cell to have high forward current values, forced by the other cells in series with it, with a reverse cell voltage having an absolute value that is lower than the absolute value of V_{br}. The effect of the shunt resistance is evident in Figure 6.3.

The exponent m affects the curvature of the I–V curve when it approaches V_{br}: the larger m the more gradual the current growth approaching the breakdown voltage.

Much more intuitively, α defines the weight of the breakdown current with respect to the term $\frac{V+I \cdot R_s}{R_h}$. Typical values for α and m are $\alpha = 0.002$ and $m = 3$. It is worth noting that cell producers do not give these values in cell datasheets, but only give a suggestion for the maximum number of cells connected in series that must be protected by one bypass diode. There are some differences between so-called "type A" and "type B" cells, in terms of their behavior in the second quadrant, with a sharper curve and a smaller breakdown voltage for type B [2].

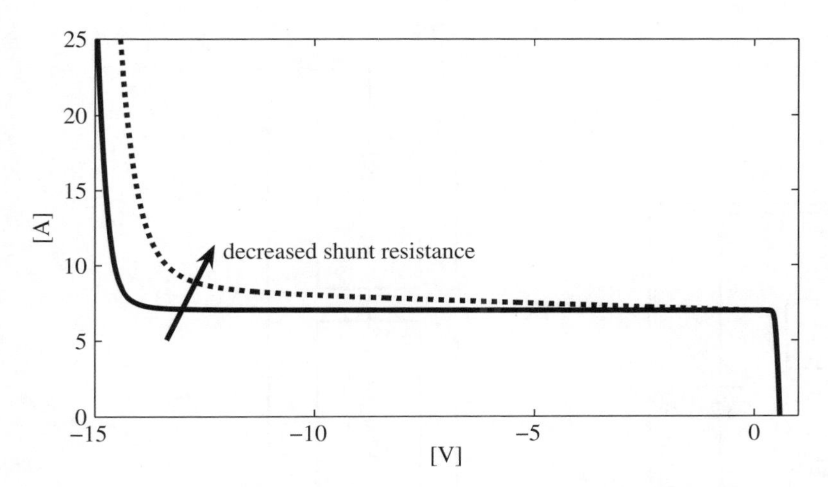

Figure 6.3 Effect of the shunt resistance on the I–V curve of a PV cell. Parameter values as per Figure 6.2, and with R_{sh} = 200 Ω (continuous line) and with R_{sh} = 10 Ω (dashed line).

Equation 6.1 shows that the I–V relationship obtained through the SDM becomes even more complicated due to the presence of the Bishop term. Nevertheless, it is still possible to work on symbolic expressions for making the solution of the non-linear model of the whole PV field more efficient. Indeed, starting from the suggestion given by Bishop [1], it is possible to use the diode voltage $V_j = V + I \cdot R_s$ shown in Figure 6.1 to get an explicit expression of the cell current I as a function of V_j, as in (6.2):

$$I = I_{ph} - I_s \cdot \left(e^{\frac{V_j}{\eta V_t}} - 1 \right) - \frac{V_j}{R_h} \cdot \left[1 + \alpha \cdot \left(1 - \frac{V_j}{V_{br}} \right)^{-m} \right] \tag{6.2}$$

thus keeping the series resistance R_s out of the non-linear relationship between the voltage and the current.

Orozco-Gutierrez et al. [3] represented a string of PV cells as in Figure 6.4. This is useful for a symbolic analysis of the system of non-linear equations describing the behavior of even a large PV field modeled down to the cell level.

6.1.2 Silicon Cells Type and Reverse Behavior

The approach proposed by Spirito and Abergamo [4] is based on a classification into cell types A and B. The former type has a cell model that does not include the shunt resistance:

$$I = \left(I_{ph} - I_s \cdot \left(e^{\frac{V}{\eta V_t}} - 1 \right) \right) \cdot M(V) \tag{6.3}$$

where M is the following function of the cell voltage, which is the Miller formula:

$$M(V) = \frac{1}{1 - \left(\frac{|V|}{V_{br}} \right)^m} \tag{6.4}$$

In (6.4), V_{br} is the junction breakdown voltage and m is the Miller exponent.

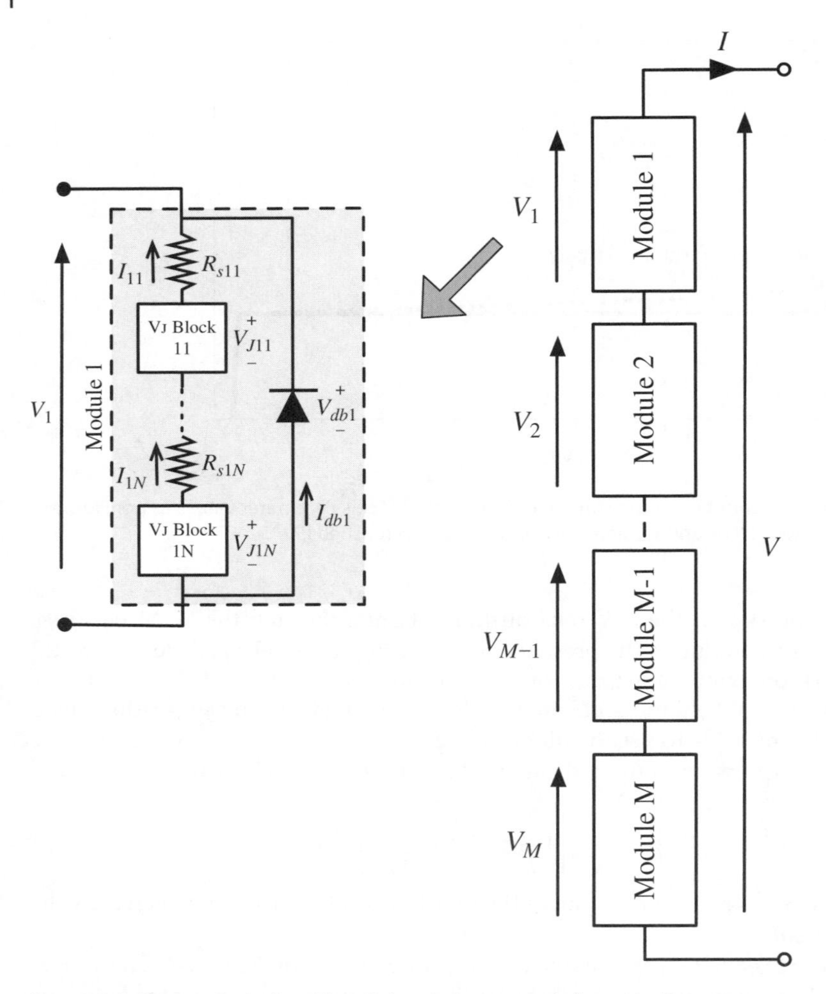

Figure 6.4 PV array including *M* modules, each one made from *N* cells.

Meanwhile, type B cells do not need the *M* term, because their reverse behavior is well modeled by the shunt resistance alone.

6.2 Cell and Subcell Modeling: Occurrence of Hot Spots

When the PV string is simulated at a module level, as shown using the approach of Chapter 5, it is assumed that the irradiation level received by all the cells in the module is the same, and also that those cells are perfectly equal in terms of parameters. Thus, if the module is driven towards reverse behavior, the bypass diode dedicated to the module enters into conduction. This means that the second-quadrant behavior of the PV string is replaced by the bypass diode forward curve. Unfortunately, this modeling level does not allow the reverse behavior of the cells to be examined; it is perhaps the best compromise between accuracy and computation speed, but it does not account for the hot-spot phenomena that the cells might be subject to in some operating conditions.

6.2.1 Cell Modeling

The cell model must therefore be improved so that it takes into account cell behavior at negative voltage values (see Equation 6.1) [1]. In this case, the PV string is modeled as shown in Figure 6.4: in the sequel, the way in which the system of non-linear equations can be written is shown. Again, as in Chapter 5, the careful way of arranging the equations in the system gives a special structure to the Jacobian matrix. Also in this case, an explicit expression of the inverse of the Jacobian matrix can be achieved and the accurate solution of the non-linear system is obtained with low computational effort.

As shown in Figure 6.4, the string consists of M modules, each having its own bypass diode, and N being the number of cells per module. The unknown variables are:

- the cells' internal voltages $V_{J_{kn}}$, the first index k being the number of modules, and the second index n being the cells of the kth module
- the bypass diodes' voltages V_{db_k}.

In each module, the sum of all the cells' voltages is equal to the voltage across their bypass diode. For example, for the mth module it is:

$$\sum_{i=1}^{N} V_{J_{mi}} - \sum_{i=1}^{N} R_{S_{Mmi}} I_{m1}(V_{J_{m1}}) = V_{db_m} \tag{6.5}$$

The cells in the module share the current, so for the mth module the current is:

$$I_{m1}(V_{J_{m1}}) = I_{m2}(V_{J_{m2}}) = I_{m3}(V_{J_{m3}}) \cdots = I_{mN}(V_{J_{mN}}) \tag{6.6}$$

Finally, the series-connected modules also share the current, so that for the $(m-1)$th and for the mth modules it is:

$$I_{m1}(V_{J_{m1}}) + I_{db_m}(V_{db_m}) = I_{(m-1)1}(V_{J_{(m-1)1}}) + I_{db_{(m-1)}}(V_{db_{(m-1)}}) \tag{6.7}$$

As for the approach described in Chapter 5, the Jacobian matrix assumes a different structure depending on the way in which the equations are written and organized. Some suitable choices must be made; the equations concerning each module are consecutive, and arranged in the following order:

- the Kirchhoff voltage law concerning the cells and the bypass diode are first
- the Kirchhoff current laws concerning the current in the first cell of the module that is related with the others are put just after
- then the Kirchhoff current law relating one module with the next appears.

The equations are listed by organizing the modules in descending order, as in Figure 6.4, so that the $N+1$ equations concerning the Mth module appear first, followed by those referring to the $(M-1)$th module and so on. The system ends with the Kirchhoff voltage law collecting the modules' voltages and the PV string voltage. Thus the non-linear system consists of $M \cdot (N+1)$ equations. The resulting system $F(X)$ is as given in (6.8), where the vector of unknown variables is: $X = [V_{J_{11}}, \dots, V_{J_{1N}}, \dots, V_{J_{M1}}, \dots, V_{J_{MN}}, V_{db_1}, \dots, V_{db_M}]$.

$$F(X) = \begin{cases} \sum_{i=1}^{N} V_{J_{Mi}} - \sum_{i=1}^{N} R_{s_{Mi}} I_{M1}(V_{J_{M1}}) - V_{db_M} = 0 \\ I_{M1}(V_{J_{M1}}) - I_{M2}(V_{J_{M2}}) = 0 \\ \vdots \\ I_{M1}(V_{J_{M1}}) - I_{MN}(V_{J_{MN}}) = 0 \\ I_{M1}(V_{J_{M1}}) + I_{db_M}(V_{db_M}) - I_{(M-1)1}(V_{J_{(M-1)1}}) - I_{db_{(M-1)}}(V_{db_{(M-1)}}) = 0 \\ \sum_{i=1}^{N} V_{J_{(M-1)i}} - \sum_{i=1}^{N} R_{s_{(M-1)i}} I_{(M-1)1}(V_{J_{(M-1)1}}) - V_{db_{(M-1)}} = 0 \\ I_{(M-1)1}(V_{J_{(M-1)1}}) - I_{(M-1)2}(V_{J_{(M-1)2}}) = 0 \\ \vdots \\ I_{(M-1)1}(V_{J_{(M-1)1}}) - I_{(M-1)N}(V_{J_{(M-1)N}}) = 0 \\ I_{(M-1)1}(V_{J_{(M-1)1}}) + I_{db_{(M-1)}}(V_{db_{(M-1)}}) - I_{(M-2)1}(V_{J_{(M-2)1}}) - I_{db_{(M-2)}}(V_{db_{(M-2)}}) = 0 \\ \vdots \\ \sum_{i=1}^{N} V_{J_{1i}} - \sum_{i=1}^{N} R_{s_{1i}} I_{11}(V_{J_{11}}) - V_{db_1} = 0 \\ I_{11}(V_{J_{11}}) - I_{12}(V_{J_{12}}) = 0 \\ \vdots \\ I_{11}(V_{J_{11}}) - I_{1N}(V_{J_{1N}}) = 0 \\ V_{db_1} + V_{db_2} + \cdots + V_{db_M} - V = 0 \end{cases} \tag{6.8}$$

Once again, the non-linear behavior of the bypass diodes is modeled using the Shockley equation:

$$I_{db_k} = I_{s,db_k}(e^{V_{db_k}/n_{db_k} V_{t,db_k}} - 1) \tag{6.9}$$

This organization of the system of equations leads to a Jacobian matrix J that is sparse. The matrix J is obtained using the differential impedance of the cells and of the bypass diodes: R_{kn} is the differential resistance of the nth cell in the kth module (see Equation 6.10). Similarly, R_{db_k}, appearing in (6.11), is the differential resistance of the bypass diode connected in anti-parallel with the kth module. The sum of all the N series resistances in the kth module is $\sum_{i=1}^{N} R_{s_{ki}}$. Finally, the Jacobian matrix J is expressed as shown in Matrix 6.1 on page 119.

$$\frac{1}{R_{kn}} = \frac{\partial I_{kn}}{\partial V_{J_{kn}}} \tag{6.10}$$

$$\frac{1}{R_{db_k}} = \frac{\partial I_{db_k}}{\partial V_{db_k}} \tag{6.11}$$

In order to apply the same reasoning, based on the Schür complement, which has already been used fruitfully in Chapter 5, the Jacobian matrix J is divided into smaller blocks. In a recursive way, it is partitioned into further submatrices as shown in Figure 6.5. The Jacobian J is divided into $M-1$ clusters J_k and each cluster k is in turn divided into four submatrices A_k, B_k, C_k, and D_k.

$$
\begin{pmatrix}
1-\frac{\sum_{i=1}^{N} R_{s_{Mi}}}{R_{M1}} & 1 & \cdots & 1 & -1 & 0 & 0 & 0 & 0 & 0 & \vdots & 0 & 0 & 0 & 0 & 0 \\
\frac{1}{R_{M1}} & -\frac{1}{R_{M2}} & 0 & 0 & 0 & 0 & 0 & 0 & 0 & 0 & \vdots & 0 & 0 & 0 & 0 & 0 \\
\vdots & 0 & \ddots & 0 & 0 & 0 & 0 & 0 & 0 & 0 & \vdots & 0 & 0 & 0 & 0 & 0 \\
\frac{1}{R_{M1}} & 0 & 0 & -\frac{1}{R_{MN}} & 0 & 0 & 0 & 0 & 0 & 0 & \vdots & 0 & 0 & 0 & 0 & 0 \\
\frac{1}{R_{M1}} & 0 & 0 & 0 & \frac{1}{R_{db_M}} & \frac{-1}{R_{(M-1)1}} & 0 & 0 & 0 & \frac{-1}{R_{db_{(M-1)}}} & \vdots & 0 & 0 & 0 & 0 & 0 \\
0 & 0 & 0 & 0 & 0 & 1-\frac{\sum_{i=1}^{N} R_{s_{(M-1)i}}}{R_{(M-1)1}} & 1 & \cdots & 1 & -1 & \vdots & 0 & 0 & 0 & 0 & 0 \\
0 & 0 & 0 & 0 & 0 & \frac{1}{R_{(M-1)1}} & -\frac{1}{R_{(M-1)2}} & 0 & 0 & 0 & \vdots & 0 & 0 & 0 & 0 & 0 \\
0 & 0 & 0 & 0 & 0 & \vdots & 0 & \ddots & 0 & 0 & \vdots & 0 & 0 & 0 & 0 & 0 \\
0 & 0 & 0 & 0 & 0 & \frac{1}{R_{(M-1)1}} & 0 & 0 & -\frac{1}{R_{(M-1)N}} & 0 & \vdots & 0 & 0 & 0 & 0 & 0 \\
0 & 0 & 0 & 0 & 0 & \frac{1}{R_{(M-1)1}} & 0 & 0 & 0 & \frac{1}{R_{db_{(M-1)}}} & \vdots & -\frac{1}{R_{11}} & 0 & 0 & 0 & -\frac{1}{R_{db_1}} \\
0 & 0 & 0 & 0 & 0 & 0 & 0 & 0 & 0 & 0 & \ddots & 1-\frac{\sum_{i=1}^{N} R_{s_{1i}}}{R_{11}} & 1 & \cdots & 1 & -1 \\
0 & 0 & 0 & 0 & 0 & 0 & 0 & 0 & 0 & 0 & \vdots & \frac{1}{R_{11}} & -\frac{1}{R_{12}} & 0 & 0 & 0 \\
0 & 0 & 0 & 0 & 0 & 0 & 0 & 0 & 0 & 0 & \vdots & \vdots & 0 & \ddots & 0 & 0 \\
0 & 0 & 0 & 0 & 0 & 0 & 0 & 0 & 0 & 0 & \vdots & \frac{1}{R_{11}} & 0 & 0 & -\frac{1}{R_{1N}} & 0 \\
0 & 0 & 0 & 0 & 1 & 0 & 0 & 0 & 0 & 1 & \vdots & 0 & 0 & 0 & 0 & 1
\end{pmatrix}
$$

Matrix 6.1 Jacobian matrix for the system of equations describing the string behavior at cell level.

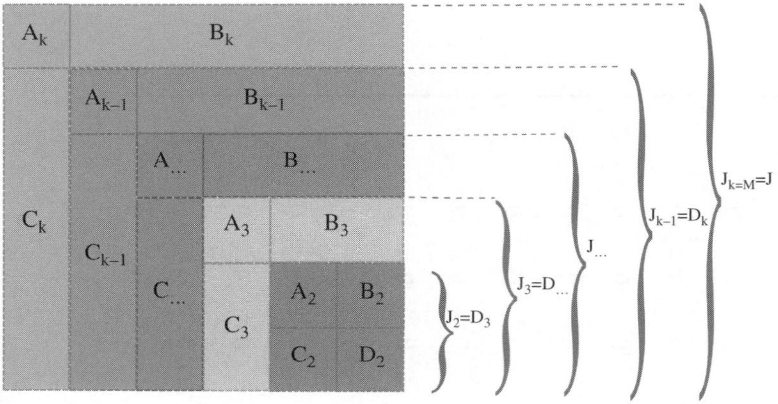

Figure 6.5 Clustering of Jacobian matrix J.

In such a structure, $D_k = J_{k-1}$ is the inner cluster that is subdivided again. The submatrices have a size that can be easily determined: A_k is a square matrix of size $(N+1)$, B_k has a size of $(N+1) \times (N+1)(k-1)$, C_k has $(N+1)(k-1)$ rows and $(N+1)$ columns and D_k is a square matrix of size $(N+1)(k-1)$.

The submatrices A_k have the following general structure:

$$\mathbf{A}_k = \begin{pmatrix} 1 - \frac{\sum_{i=1}^{N} R_{s_{ki}}}{R_{k1}} & 1 & \cdots & 1 & -1 \\ \frac{1}{R_{k1}} & \frac{-1}{R_{k2}} & 0 & 0 & 0 \\ \vdots & 0 & \ddots & 0 & 0 \\ \frac{1}{R_{k1}} & 0 & 0 & \frac{-1}{R_{kN}} & 0 \\ \frac{1}{R_{k1}} & 0 & 0 & 0 & \frac{1}{R_{db_k}} \end{pmatrix} \tag{6.12}$$

The submatrices B_k always exhibit the same number of rows, while the number of columns is variable: the first $N+1$ are invariant, and a variable number of columns full of zeros, depending on the index k, is added.

$$\mathbf{B}_k = \left[\underbrace{\begin{matrix} 0 & 0 \cdots & 0 & 0 \\ 0 & 0 \cdots & 0 & 0 \\ 0 & 0 \cdots & 0 & 0 \\ 0 & 0 \cdots & 0 & 0 \\ -\frac{1}{R_{(k-1)1}} & 0 \cdots & 0 & -\frac{1}{R_{db_{(k-1)}}} \end{matrix}}_{\text{Fixed part } (N+1)\times(N+1)} \begin{matrix} | \\ | \\ | \\ | \\ | \end{matrix} \underbrace{\begin{matrix} 0 & 0 \cdots & 0 & 0 \\ 0 & 0 \cdots & 0 & 0 \\ 0 & 0 \cdots & 0 & 0 \\ 0 & 0 \cdots & 0 & 0 \\ 0 & 0 \cdots & 0 & 0 \end{matrix}}_{\text{Variable part } (N+1)\times(N+1)(k-2)} \right] \tag{6.13}$$

In the same way, the submatrices C_k have a fixed number of columns and a variable number of rows: the upper part is full of zeros and the lower part has a fixed number of non-zero rows.

$$\mathbf{C}_k = \begin{bmatrix} 0 & 0 & 0 & 0 & 0 \\ 0 & 0 & 0 & 0 & 0 \\ \vdots & \vdots & \vdots & \vdots & \vdots \\ 0 & 0 & 0 & 0 & 0 \\ 0 & 0 & 0 & 0 & 0 \\ - & - & - & - & - \\ 0 & 0 & 0 & 0 & 0 \\ 0 & 0 & 0 & 0 & 0 \\ \vdots & \vdots & \vdots & \vdots & \vdots \\ 0 & 0 & 0 & 0 & 0 \\ 0 & 0 & 0 & 0 & 1 \end{bmatrix} \begin{matrix} \left.\vphantom{\begin{matrix}0\\0\\0\\0\\0\end{matrix}}\right\} \text{Variable part}(N+1)(k-2)\times(N+1) \\ \\ \left.\vphantom{\begin{matrix}0\\0\\0\\0\\0\end{matrix}}\right\} \text{Fixed part}(N+1)\times(N+1) \end{matrix} \tag{6.14}$$

The clustering process runs up to $k = 2$, where $D_2 = A_1$. The matrix A_1 has an expression that is different from the other A_k, which are obtained using (6.12). This is:

$$\mathbf{D}_2 = \mathbf{A}_1 = \begin{pmatrix} 1 - \frac{\sum_{i=1}^{N} R_{s1i}}{R_{11}} & 1 & \cdots & 1 & -1 \\ \frac{1}{R_{11}} & \frac{-1}{R_{12}} & 0 & 0 & 0 \\ \vdots & 0 & \ddots & 0 & 0 \\ \frac{1}{R_{11}} & 0 & 0 & \frac{-1}{R_{1N}} & 0 \\ 0 & 0 & 0 & 0 & 1 \end{pmatrix} \tag{6.15}$$

For $k = 3 \ldots M$, the inverse matrices D_k^{-1} are:

$$\mathbf{D}_k^{-1} = \mathbf{J}_{k-1}^{-1} \tag{6.16}$$

In conclusion, once the size of the PV string to be simulated has been defined – the number M of modules and the number N of cells per module – the clusters and the sub-matrices are built up using (6.10) and (6.11), and the Jacobian matrix (6.1) is calculated.

The structure of the Jacobian matrix, which is the result of the process described above, allows the Schür complement to be used in order to calculate the inverse Jacobian matrix J^{-1}. It is obtained by using the Schür complement iteratively [5]. This mathematical tool allows the inverse matrix of each cluster k of J to be calculated, starting from the lower right-hand corner of J: from $k = 2$ up to $k = M$. Then, taking into account that $D_k^{-1} = J_{k-1}^{-1}$, the Schür complement allows all the J_k^{-1} from A_k, B_k, C_k, and D_k^{-1} to be calculated, as was done in Chapter 5 using (5.23), with each S_k evaluated through (5.22).

The calculation of each J_k^{-1} requires the inverse of S_k and D_k. Profiting from the sparsity of A_k, C_k, and B_k, the inverse of S_k is determined. This is shown in detail by Orozco-Gutierrez et al. [3]. Some simulation examples presented in the same paper are briefly outlined in the next section.

6.3 Simulation Example

A PV array consisting of two modules connected in series is considered. Each one includes 20 cells: all the cells in the first and in the second module work at

Table 6.1 Cell and bypass diode parameters used.

Parameter	Value
I_{ph} [A]	3.798
I_s [A]	1.26×10^{-9}
η_d [-]	1.5
R_s [Ω]	1×10^{-3}
R_h [Ω]	1000
V_{br} [V]	−15
m [-]	3
α [-]	0.002
$I_{s_{db}}$ [A]	5.6×10^{-6}
η_{db} [-]	1.5
T [°C]	25

$G = 1000\,\text{W/m}^2$, except for one in the second module working at $G = 100\,\text{W/m}^2$, being subject to shadowing. The cell parameters are assumed to be exactly equal. Table 6.1 gives the array parameter values used in the simulation.

The I–V curve resulting from the simulation is shown in Figure 6.6: the PV array current as well as the currents of the first and the second modules are shown. In addition, the I–V curve of the shadowed cell and un-shadowed cells of the second module are given. The simulation reveals that the MPP of the PV array is 50.6 W, with the shadowed cell dissipating 52.7 W and the un-shadowed cells providing 51.3 W. Module 1 generates the remaining power. At this operating point the shadowed cell is in breakdown,

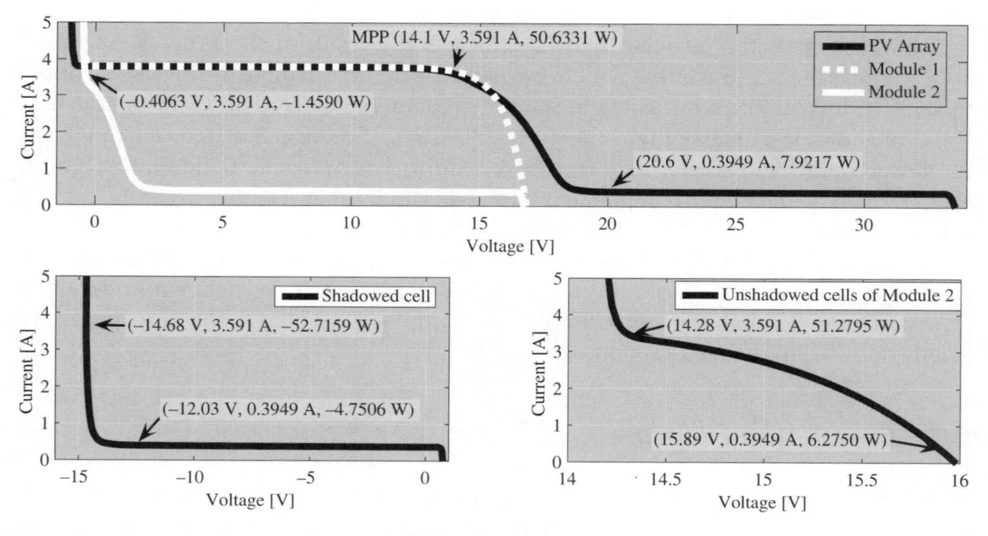

Figure 6.6 I–V curve for PV system with a shadowed cell.

as shown in Figure 6.6. In order to avoid this situation, the PV array's operating point should change such that the reverse voltage of the shadowed cell does not exceed a security factor, which could be close to 80% of its breakdown voltage. In this case the PV array power drops to about 7.9 W.

6.4 Subcell PV Model

The subcell model should be useful for lowering the granularity level of the representation, thus allowing it to describe if a single cell is affected by a partial shadowing or not. The objective is to achieve a simulation of such operating conditions by always using an equivalent circuit model, thus still using concentrated parameters and allowing a reasonably fast simulation. The subcell model would be useful to investigate in much more detail the effects of small shadows on the electrical and thermal stresses that the cells would be subject to in these conditions.

Few attempts to do this have been presented in the literature. Dongaonkar et al. [6] use a bi-dimensional simulation of thin film PV modules, trying to predict the uneven distribution of currents and voltages due to partial shading. Each module includes a number of series-connected cells, each decomposed into a number of parallel-connected subcells. The latter are connected to each other through the contact sheet resistance of the top and bottom contacts, so that the series and parallel subcell resistances are obtained by scaling the sheet resistance of the contact materials by the subcell aspect ratio, which is the ratio between the shadowed and the un-shadowed areas. The simulations are conducted by assuming that the partially shaded regions receive diffused light only: 20% of the direct sunlight received by the fully illuminated areas. When compared to experimental results, the proposed model appears to be reliable.

A similar approach was introduced by Olalla et al. for crystalline silicon cells [7]. In particular, in the presence of a shading factor:

$$\alpha = \frac{\text{shadowed area}}{\text{total area}} \in [0, 1] \tag{6.17}$$

The photo-induced current at which the cell works is calculated as a function of the shading factor (6.17) and of G and D, the global and diffuse irradiance, respectively.

$$I_{ph} = \frac{I_{ph,STC}}{G_{STC}} \cdot (G \cdot (1 - \alpha) + D \cdot \alpha) \tag{6.18}$$

In order to get a fast simulation of the PV array, the authors approximate the non-linear behavior of each PV cell and of the bypass diodes by means of a piecewise linear function. The voltage range is partitioned into a number of intervals in which the device behaves linearly: the extremes of the voltage intervals as well as the slope of the curve in each interval are preliminarily computed and stored in lookup tables. In this way, during the simulation of the complete PV system, instead of calculating the values of a non-linear function, the lookup tables are accessed, and this speeds the computation up. Moreover, the slopes of the curves, saved in the lookup tables, are used for the Jacobian matrix, which is useful for accelerating the solution of the non-linear system describing the PV generator. The authors also show the effectiveness of the approach in long-term simulations aimed at assessing the energy production of a PV power plant.

In the work of Wu et al. [8], the SDM of the PV source remains the core of the simulation tool, but a hierarchical architecture allows the mismatching effects to be reproduced with a computational load that is lower than that required by a finite element model. Resistances are used to take into account the ohmic effects related to the front and back contacts of the basic element of the model, namely the subcell. Nodal analysis is used for writing the equations of the lumped network describing the cells as aggregations of subcells and, again, the Newton–Raphson approach is used to solve the resulting system of non-linear equations.

In addition to the previously described methods, an approach similar to that described in Chapter 5 for module-based simulation can be used. In particular, the shadowed and un-shadowed portions of each PV cell can be described by the parallel connection of two SDMs. The two circuits share the same cell voltage but contribute with a current that depends on the parameter values and working conditions of the subportion of the cell they model. The set of parameters of the two cell sections – $\{I_{s,h}, V_{t,h}, I_{ph,h}, R_{s,h}, R_{h,h}\}$ for the un-shadowed section and $\{I_{s,l}, V_{t,l}, I_{ph,l}, R_{s,l}, R_{h,l}\}$ for the shadowed section – are determined using the procedure described in Chapter 2 and by accounting for the respective surface areas A_h and A_l and the values of the irradiance and temperature for the two cell portions $\{G_h, T_h\}$ and $\{G_l, T_l\}$.

6.5 Concluding Remarks on PV String Modeling

At the end of this chapter it is worth giving some conclusions to the reader about PV string modeling.

Chapter 3 has given readers the tools for simulating PV arrays in uniform conditions. The identification of the circuit model parameters is done using the approaches presented in Chapter 2 as well as the translation of the values identified for STC to any other working condition. The presence of mismatching effects, regardless of their origin, requires more powerful simulation methods. The computation and memory loads grow significantly, and a careful choice of an appropriate model becomes mandatory. Chapter 5 gives an approach at module level, which assumes that all the cells in a model are not subject to any mismatching effect. This is of course an approximation, but it represents a good compromise between the computational load and the accuracy of the result. A more detailed simulation requires an approach that is at cell level and which also accounts for the reverse voltage behavior of the cells, as described in this chapter. The consequent computational burden is obviously higher than that for the previous methods.

References

1 Bishop, J. (1988) Computer simulation of the effects of electrical mismatches in photovoltaic cell interconnection circuits. *Solar Cells*, **25**, 73–89.

2 Molenbroek, E., Waddington, D., and Emery, K. (1991) Hot spot susceptibility and testing of pv modules, in *Photovoltaic Specialists Conference, 1991, Conference Record of the Twenty Second IEEE*, pp. 547–552, vol.1, doi:10.1109/PVSC.1991.169273.

3 Orozco-Gutierrez, M., Ramirez-Scarpetta, J., Spagnuolo, G., and Ramos-Paja, C. (2014) A method for simulating large PV arrays that include reverse biased cells. *Applied Energy*, **123** (0), 157–167, doi:10.1016/j.apenergy.2014.02.052.

4 Spirito, P. and Abergamo, V. (1982) Reverse bias power dissipation of shadowed or faulty cells in different array configurations, in *Proceedings of the Fourth European Photovoltaic Solar Energy Conference*, pp. 296–300.

5 Zhang, F. The Schur complement and its applications – numerical methods and algorithms.

6 Dongaonkar, S., Deline, C., and Alam, M. (2013) Performance and reliability implications of two-dimensional shading in monolithic thin-film photovoltaic modules. *Photovoltaics, IEEE Journal of*, **3** (4), 1367–1375, doi:10.1109/JPHOTOV.2013.2270349.

7 Olalla, C., Clement, D., Maksimovic, D., and Deline, C. (2013) A cell-level photovoltaic model for high-granularity simulations, in *Power Electronics and Applications (EPE), 2013 15th European Conference on*, pp. 1–10, doi:10.1109/EPE.2013.6631946.

8 Wu, X., Bliss, M., Sinha, A., Betts, T., Gupta, R., and Gottschalg, R. (2014) Distributed electrical network modelling approach for spatially resolved characterisation of photovoltaic modules. *Renewable Power Generation, IET*, **8** (5), 459–466, doi:10.1049/iet-rpg.2013.0242.

7

Modeling the PV Power Conversion Chain

7.1 Introduction

The current delivered by a PV source depends on the environmental conditions and operating voltage, as illustrated in the previous chapters. This aspect has been extensively discussed in the literature, mainly focusing on the tracking of the optimal operation voltage. That tracking is usually implemented by connecting a DC/DC switching converter between the PV source and a battery or a load, to set the optimal PV voltage or load to the source terminals. Therefore, it is expected that in the near future commercial PV panels will be sold with integrated DC/DC converters, so the PV source will be formed by both the modules and the converter.

The DC/DC converter influences the system performance through three effects:

- steady-state losses and voltage transformations that modify the power-vs-voltage curve of the overall system that interacts with the load
- PV voltage and current ripples
- transients at the source terminals due the converter dynamics.

The first effect is related to the efficiency and voltage-conversion ratio of the DC/DC converter. Since the efficiency of the DC/DC converter is, in general, a non-linear function of the processed power [1], the operating point at which the source produces the maximum power does not necessarily match the operating point at which the load receives the maximum power. The PV power conversion chain formed by the PV source and the DC/DC converter must therefore be modeled to detect the optimal operation voltage, at which the maximum power is transferred to the load. Such a model is useful for design and test of maximum power point tracking (MPPT) algorithms in realistic conditions and to accurately predict the energy production of a PV installation.

The second effect is the high-frequency ripple produced by the switching operation of the DC/DC converter at the source terminals. This ripple prevents the PV operating point from being kept at the MPP, thus reducing the power production. The use of large capacitors and/or inductors in the DC/DC converter allows the ripple amplitude to be reduced, but this approach increases the converter cost and weight. Therefore, it is necessary to model the relationship between the high-frequency ripples, the converter components, and the PV source to reach a trade-off between voltage-ripple magnitude and converter size, weight, and cost.

The third effect is related to the low-frequency dynamics introduced by the capacitors and inductors of the DC/DC converter in the PV power-conversion chain. These

Photovoltaic Sources Modeling, First Edition. Giovanni Petrone, Carlos Andrés Ramos-Paja and Giovanni Spagnuolo.

Companion Website: www.wiley.com/go/petrone/Photovoltaic_Sources_Modeling

dynamics introduce delays between the command provided by an MPPT algorithm and the PV voltage, which could make the MPPT control loop unstable. Moreover, changes in the environmental and load conditions affect the DC/DC converter operating point, which in turn affects the PV power production. Therefore, it is necessary to model and control the dynamic behavior of the whole PV power conversion chain to ensure a stable operation.

This chapter examines these effects from the photovoltaic source point of view. The analyses are illustrated by considering module integrated units (MIUs), which are composed of a single PV module and the associated DC/DC converter. The output terminals of the converter are connected to a DC link, which is represented by means of a voltage source. Such a modular PV system, depicted in Figure 7.1, is widely used by manufacturers of PV inverters to provide commercial distributed MPPT (or DMPPT) solutions. One example is the Power Optimizer from SolarEdge [2]. The scheme in Figure 7.1 shows the presence of a voltage source at the DC/DC converter output terminals. This models the common case of connection at a DC bus. The addition of a resistor in series with the DC source, allows to analyze the case of a PV battery charger, and can be studied in a similar way to that outlined for the scheme in Figure 7.1. The operation of a MIU involves few variables in comparison with large PV arrays, so the illustration of the proposed analyses is simpler, which makes the examples introduced in this chapter easy to understand. The analyses provided in this chapter can be extended to larger PV sources, such as strings or arrays, by scaling up the PV source model.

Commercial MIUs typically operate with an MPPT algorithm acting directly on the duty cycle of the DC/DC converter, but in some conditions its performance is degraded by environmental and load perturbations, for example when there are voltage oscillations caused by the grid connection. A controller could therefore be added to regulate the DC/DC converter, to mitigate the effects of both environmental conditions and load perturbations, and at the same time to follow the commands of the MPPT algorithm with the desired promptness.

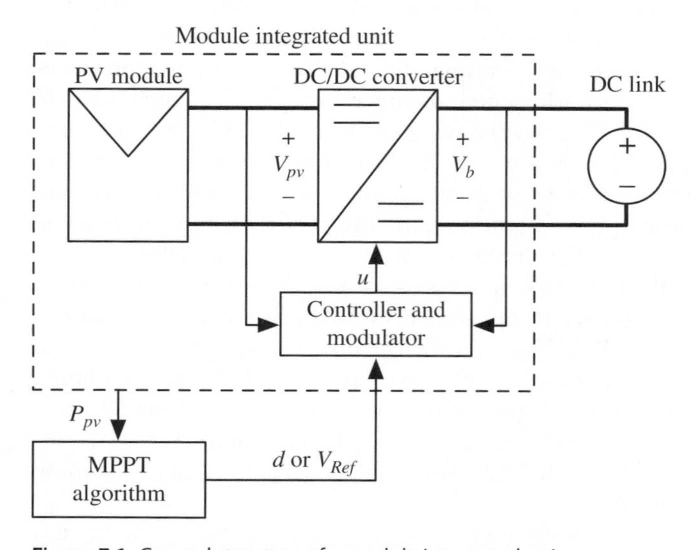

Figure 7.1 General structure of a module integrated unit.

Section 7.3 examines the effect of the DC/DC converter on the steady-state operation of the MIU. The analyses are performed using classical approaches for DC/DC converter analysis, which are reviewed in Section 7.2. Then, Section 7.4 looks at the dynamic behavior of the MIU. Section 7.5 introduces additional examples concerning MIUs, based on different DC/DC converters and covering boost, buck, and buck–boost topologies. Finally, Section 7.6 gives a summary of the chapter.

7.2 Review of Basic Concepts for Modeling Power Converters

Power converters are commonly analyzed by averaging the current and voltage waveforms within the switching period. Such a technique makes it possible to simplify the converter analysis by neglecting the switching ripple produced by the converter operation. This section reviews the main concepts required to model power converters, while a complete description of the averaging modeling technique is available in the literature [1, 3].

The operation of power converters involves the commutation of a number of semiconductor switches to form different topologies. Therefore, power converters are also known as circuits with variable structure. A first example of that behavior is given in Figure 7.2, where a classical lossless boost converter is presented. Such a circuit is composed by an inductor L, a capacitor C, a MOSFET *Mos* and a diode D_d. In addition, a power source V_g and the load R are also connected to the converter input–output ports.

The circuit operates by changing the states of both the MOSFET and the diode between ON and OFF. When the MOSFET is set ON, the diode is automatically set OFF, because its voltage becomes negative. Similarly, when the MOSFET is set OFF, the diode is set ON. Such behavior is controlled using the gate voltage u of the MOSFET, which can assume normalized values 0 (*Mos* OFF) and 1 (*Mos* ON).

Figure 7.2 illustrates the periodic waveform of u, where the period between two consecutive MOSFET activations is denoted by T_{sw}. In addition, Figure 7.2 shows the two

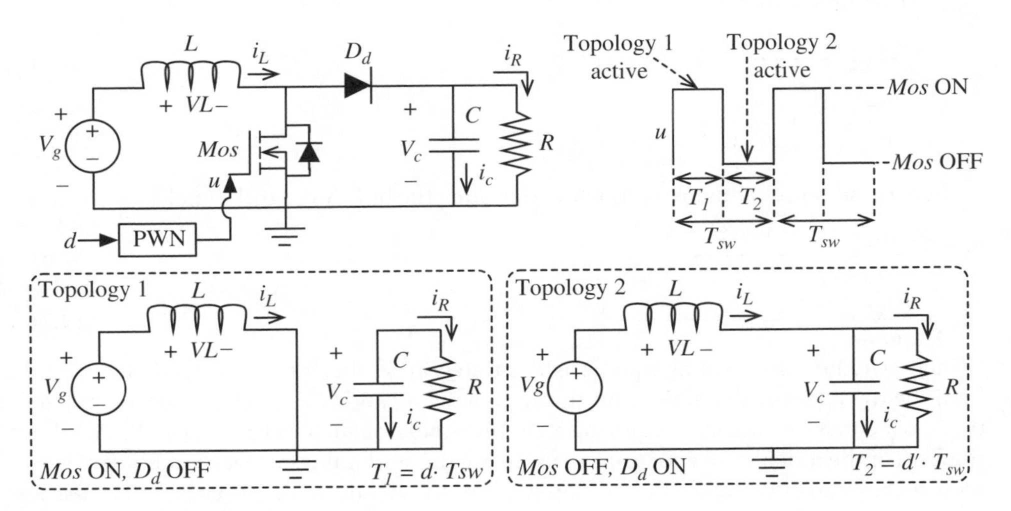

Figure 7.2 Boost converter operation.

circuital topologies generated by the MOSFET operation. Topology 1 occurs when *Mos* is ON and D_d is OFF; that is, when $u = 1$. Topology 2 occurs when *Mos* is OFF and D_d is ON; that is, when $u = 0$. The time interval T_1 in which topology 1 is active is described in relation with the period as $T_1 = d \cdot T_{sw}$, where $0 < d < 1$ is the converter duty cycle. Hence, the time interval T_2 in which topology 2 is active is given by $T_2 = d' \cdot T_{sw}$, where $d' = 1 - d$ is the complementary duty cycle.

The modeling of the power converter is performed by describing the behavior of each topology using differential equations. In such a way, the converter behavior is described in terms of its *state variables*, which commonly correspond to the inductor currents and capacitor voltages [1, 4], but other electrical or mathematical quantities can be also chosen. The inductor current i_L and capacitor voltage v_C are the states selected to model the boost converter in Figure 7.2. In this chapter, all the variables in lower case text correspond to time-varying quantities.

Next, the differential equations of each state are described in terms of u, to model the transition between the different topologies. Such a procedure is described in (7.1), where the derivatives of the state vector **X** are expressed as functions of the states, inputs **U**, and time t.

$$\frac{\partial \mathbf{X}}{\partial t} = \mathbf{F}(\mathbf{X}, \mathbf{U}, t) \tag{7.1}$$

In the example of the boost converter, the state vector is $\mathbf{X} = \left[i_L \; v_C \right]^T$, the input vector is $\mathbf{U} = \left[V_g \right]$, and the differential equations that describe the converter behavior in both topologies are:

Topology 1 $(u = 1)$:

$$\frac{\partial i_L}{\partial t} = \frac{V_g}{L} \tag{7.2}$$

$$\frac{\partial v_C}{\partial t} = -\frac{v_C}{R \cdot C} \tag{7.3}$$

Topology 2 $(u = 0)$:

$$\frac{\partial i_L}{\partial t} = \frac{V_g - v_C}{L} \tag{7.4}$$

$$\frac{\partial v_C}{\partial t} = \frac{i_L}{C} - \frac{v_C}{R \cdot C} \tag{7.5}$$

Then, these expressions are combined to include the binary control signal u as follows:

$$\frac{\partial i_L}{\partial t} = \frac{V_g - v_C \cdot (1 - u)}{L} \tag{7.6}$$

$$\frac{\partial v_C}{\partial t} = \frac{i_L \cdot (1 - u)}{C} - \frac{v_C}{R \cdot C} \tag{7.7}$$

Such switched differential equations accurately describe the converter behavior in both steady-state and dynamic conditions, so that (7.6) and (7.7) can be used to simulate the boost converter without requiring a circuit simulator such as PSIM or PSPICE. The solution of the differential equations to calculate the state values in each simulation step requires the use of the Euler integration method [5]. In addition, it is common to use a fixed frequency driver, based on the pulse width modulation, as shown in Figure 7.2, to impose the duty cycle d on the converter [1, 3].

To illustrate the boost converter behavior predicted by the switched differential equations (7.6) and (7.7), the following parameters are considered: $L = 100\,\mu$H, $C = 50\,\mu$F, $V_g = 10$ V, $R = 5\,\Omega$, $T_{sw} = 10\,\mu$s. In addition, the simulation starts with a duty cycle $d = 0.48$, imposed by a pulse width modulator, but at $t = 8$ ms the duty cycle is changed to $d = 0.5$ to illustrate the converter dynamic behavior. Figure 7.3 shows the simulation result obtained through the switched differential equations in black traces: the low-frequency component and a high-frequency ripple are both evidenced.

The ripple affecting the inductor current and the capacitor voltage is periodic, with period T_{sw} [1]. Since ripples may affect both sources and loads [1, 6, 7], the power converters are commonly designed to exhibit low ripple magnitudes in comparison with the steady state magnitudes. The ripple components are usually neglected (see for example Eriksson et al. [1]) and the converter is described by means of its low-frequency components only. In such a technique, known as the *small-ripple approximation* [1], the binary control signal u is replaced with its average value within the switching period, which corresponds to the duty cycle d of the converter:

$$d = \frac{1}{T_{sw}} \int_0^{T_{sw}} u \, dt \tag{7.8}$$

Then, the switched differential equations (7.1) are averaged by replacing u with d, as in (7.9), where $\overline{\mathbf{X}}$ is the vector of the averaged states. To illustrate this procedure, the averaged differential equations for the boost converter are given in (7.10) and (7.11).

$$\frac{\partial \overline{\mathbf{X}}}{\partial t} = \mathbf{F}\left(\overline{\mathbf{X}}, \mathbf{U}, t\right)\Big|_{u=d} \tag{7.9}$$

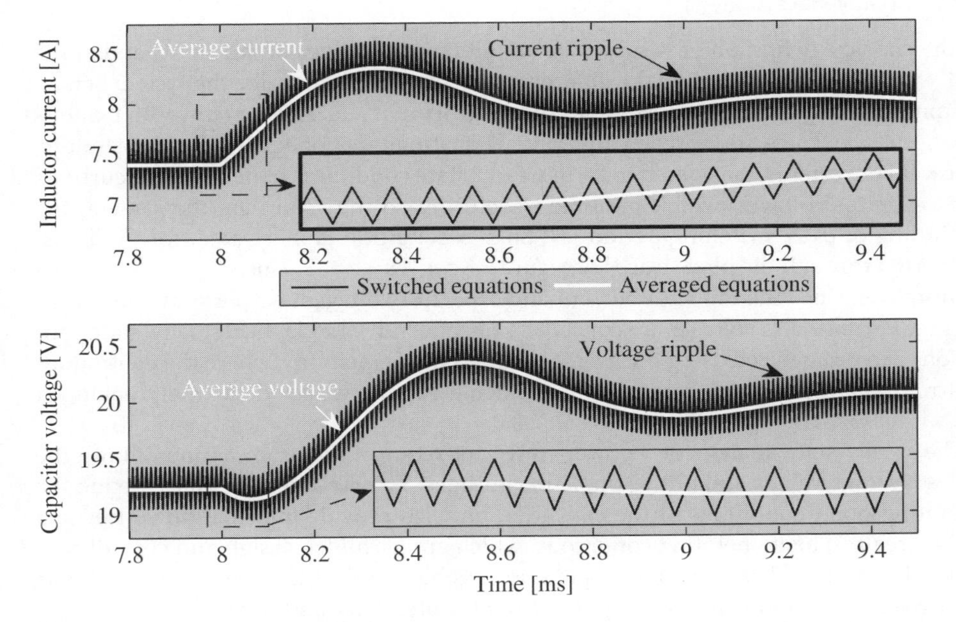

Figure 7.3 Boost converter simulation.

$$\frac{\partial \overline{i_L}}{\partial t} = \frac{V_g - \overline{v_C} \cdot (1-d)}{L} \tag{7.10}$$

$$\frac{\partial \overline{v_C}}{\partial t} = \frac{\overline{i_L} \cdot (1-d)}{C} - \frac{\overline{v_C}}{R \cdot C} \tag{7.11}$$

In such equations, the control signal d is continuous instead of binary (discontinuous). Moreover, the converter dynamics are averaged within the switching period, so the low-frequency components are modeled but the high-frequency ripple is neglected. Since power converters are commonly designed to provide small ripples, the accuracy of such averaged differential equations is acceptable. This feature is evident in Figure 7.3, where the averaged differential equations (7.10) and (7.11) give the white traces.

The continuous control signal d, which appears in the averaged equations, simplifies the analysis and control of the system [1, 4] in comparison with the switched equations, which include the discontinuous control signal u. This is illustrated in the following sections, where the averaged approach allows linear transfer functions that are used to analyze the system dynamics to be obtained. It must be pointed out that the input vector of the averaged system must include the duty cycle, so it is possible to calculate transfer functions with respect to the duty cycle. Thus the input vector of the boost converter example becomes $\mathbf{U} = \left[d\ V_g\right]^T$.

Finally, both the switched and the averaged equations allow the converter to be designed, but the averaged ones are commonly used to analyze the converter dynamics and to design classical controllers. The switched equations are used to design non-linear controllers, such as sliding-mode controllers.

The following subsections review the main concepts used in the analysis of power converters in both steady-state and dynamic conditions.

7.2.1 Steady-state Analysis

In the classical definition of steady-state conditions, the values of the inputs and of the state variables are constant, so the differential equations that describe the system behavior must be equal to zero. Instead, since the control input u of a power converter exhibits a periodic waveform, its switched differential equations are not equal to zero. Therefore, a power converter is considered to be in steady-state conditions if the inductor currents and capacitor voltages exhibit periodic waveforms, so the values are the same at the beginning of each switching period [1]. Such a condition is illustrated, for the boost converter, through the black traces of Figure 7.3 for $7.8 < t < 8.0$ ms.

Moreover, since the inductor current and capacitor voltage have periodic values, the average currents and voltages are constant. Consequently, the classical steady-state conditions can be analyzed using the averaged differential equations, where the continuous control input d is set to a constant value. Such a condition is evident when looking at the white traces of Figure 7.3 for $7.8 < t < 8.0$ ms.

The steady-state analysis of a power converter requires the examination of both the states' average values and of the ripple magnitudes. The steady-state values define the operating-point conditions of the converter, but also provide information concerning the source and load operating conditions, which are useful for design and control purposes. In addition, based on such data, it is possible to select the source most suitable for a given application and to design the circuit protections, and so on.

The ripple magnitudes are required to specify the rating of the converter elements, to analyze the power quality imposed on both the source and the load, and to analyze the converter power losses. Therefore, in the following subsections both the steady-state values and ripple magnitudes are discussed.

7.2.1.1 Steady-state Values

The steady-state values are calculated by setting the averaged differential equations equal to zero, thus solving the system given in (7.12), to obtain the values of the inductor currents, capacitor voltages, or any unknown inputs and outputs:

$$\frac{\partial \overline{\mathbf{X}}}{\partial t} = 0 \tag{7.12}$$

For example, by referring to the boost converter, the steady-state relations given in (7.13) and (7.14) are obtained. In such expressions the capital letters denote steady-state values, so V_C and I_L represent the steady-state values of $\overline{v_C}$ and $\overline{i_L}$, respectively, while D represents the steady-state value of the input variable d.

$$V_g - V_C \cdot (1 - D) = 0 \tag{7.13}$$

$$I_L \cdot (1 - D) - V_C/R = 0 \tag{7.14}$$

From the steady-state relations – e.g. (7.13) and (7.14) – it is possible to calculate the steady-state value of any state and input variable. For example, by considering the simulation results shown in Figure 7.3, the steady-state values $V_C = 19.23$ V and $I_L = 7.40$ A are calculated using (7.13) and (7.14). Such analytical results are in agreement with the simulation data for $t < 8$ ms where $D = 0.48$.

7.2.1.2 Ripple Magnitudes

The ripple is defined as the amplitude of the waveform around the steady-state value of a state variable. Ripple magnitudes are calculated from the switched differential equations: the state derivative and the interval duration are used to calculate the peak-to-peak amplitude of the state variables.

For example, the inductor current ripple in the boost converter is calculated using (7.15), referring to the first topology ($u = 1$), or using (7.16), referring to the second topology ($u = 0$).

$$\left| D \cdot T_{sw} \cdot \frac{\partial i_L}{\partial t} \right| = D \cdot T_{sw} \cdot \frac{V_g}{L} = \Delta i_L \quad \text{for} \quad u = 1 \tag{7.15}$$

$$\left| D' \cdot T_{sw} \cdot \frac{\partial i_L}{\partial t} \right| = D' \cdot T_{sw} \cdot \left(-\frac{V_g - v_C}{L} \right) = \Delta i_L \quad \text{for} \quad u = 0 \tag{7.16}$$

The previous relations are valid for continuous conduction mode (CCM), in which the inductor current is strictly greater than zero in the whole switching period. The left-hand side of Figure 7.4, shows the inductor current ripple of the boost converter operating in CCM, for which it is noted that both (7.15) and (7.16) produce the same result for Δi_L. If the inductor current in each switching period exhibits a third time interval during which it remains at zero, the converter enters a discontinuous conduction mode (DCM). The right-hand side of Figure 7.4 shows the inductor current ripple of the boost converter operating in DCM. The current ripple in DCM can be calculated from (7.15) or (7.16), by neglecting the third interval where the diode current drops to zero. Because the instant

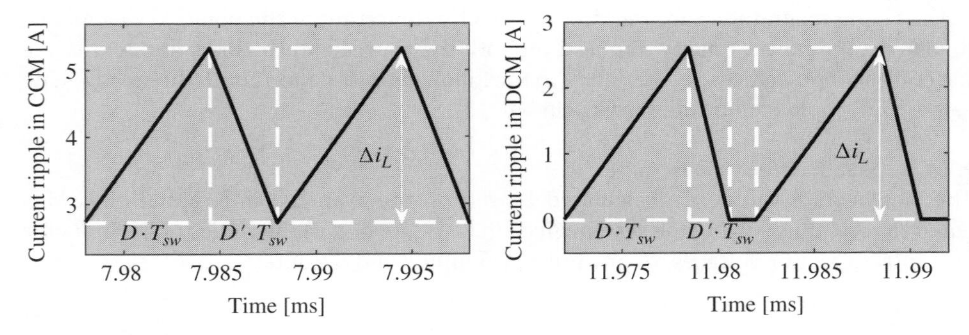

Figure 7.4 Inductor current ripple in the boost converter.

at which the diode current goes to zero requires some calculation, the DCM current ripple in the boost converter is determined from (7.15) because the first topology always considers the diode OFF.

In general, DCM is less used in industrial applications because of the load dependency of the conversion ratio, component stress, and high harmonic components produced in such an operating mode. Additional details concerning the DCM operation of a power converter are available in the literature [1].

7.2.2 Converter Dynamics Analysis

The study of converter dynamics is classically performed by linearizing the averaged differential equations to obtain transfer functions. Such a linearization process must be performed around the steady-state condition in which the converter must to operate [1, 4, 6]. Hence, the steady-state analysis of the converter is performed in advance.

The linearization process consists of evaluating the averaged equation given in (7.9) around the operating point calculated from (7.12). Then, the linearized system is used to analyze the converter dynamics using state-space models in the time domain. Moreover, it can be transformed in the Laplace domain to analyze the system poles and zeros, frequency margins, and so on [1]. Both state-space and transfer-function representations are used to design single-variable and multi-variable controls.

The first step is to construct the non-linear state-space system representing the converter. Such a system follows the structure given in (7.17) and (7.18), where $\overline{\mathbf{Y}}$ is the vector of the system outputs. Moreover, $\mathbf{A}$, $\mathbf{B}$, $\mathbf{C}$, and $\mathbf{D}$ are matrices with non-linear elements such as products of variables or time-dependent functions [6].

$$\frac{\partial \overline{\mathbf{X}}}{\partial t} = \mathbf{A} \cdot \overline{\mathbf{X}} + \mathbf{B} \cdot \mathbf{U} \tag{7.17}$$

$$\overline{\mathbf{Y}} = \mathbf{C} \cdot \overline{\mathbf{X}} + \mathbf{D} \cdot \mathbf{U} \tag{7.18}$$

The size of the system matrices depends on the number of states, inputs and outputs. In the following, N represents the number of states, M represents the number of inputs, and Q represents the number of outputs.

The matrix $\mathbf{A}$ is given in (7.19). This *states-to-states Jacobian,* describes the relation between the state's derivatives and the state's varaibles. Similarly, the matrix $\mathbf{B}$ given in (7.20) describes the relation between the state's derivatives and the inputs

$\mathbf{U} = \left[w_1, w_2, \cdots w_i \cdots w_M\right]^T$. Such a matrix is called a *states-to-inputs Jacobian*. Finally, matrices **C** and **D**, given in (7.21) and (7.22) respectively, provide the relations of the outputs with the states and inputs. Hence, **C** is called an *outputs-to-states Jacobian* and **D** is called an *outputs-to-inputs Jacobian*.

$$\mathbf{A} = \begin{bmatrix} \frac{\partial^2 x_1}{\partial t \partial x_1} & \frac{\partial^2 x_1}{\partial t \partial x_2} & \cdots & \frac{\partial^2 x_1}{\partial t \partial x_i} & \cdots & \frac{\partial^2 x_1}{\partial t \partial x_N} \\ \frac{\partial^2 x_2}{\partial t \partial x_1} & \frac{\partial^2 x_2}{\partial t \partial x_2} & \cdots & \frac{\partial^2 x_2}{\partial t \partial x_i} & \cdots & \frac{\partial^2 x_2}{\partial t \partial x_N} \\ \vdots & \vdots & \ddots & \vdots & \ddots & \vdots \\ \frac{\partial^2 x_i}{\partial t \partial x_1} & \frac{\partial^2 x_i}{\partial t \partial x_2} & \cdots & \frac{\partial^2 x_i}{\partial t \partial x_i} & \cdots & \frac{\partial^2 x_i}{\partial t \partial x_N} \\ \vdots & \vdots & \ddots & \vdots & \ddots & \vdots \\ \frac{\partial^2 x_N}{\partial t \partial x_1} & \frac{\partial^2 x_N}{\partial t \partial x_2} & \cdots & \frac{\partial^2 x_N}{\partial t \partial x_i} & \cdots & \frac{\partial^2 x_N}{\partial t \partial x_N} \end{bmatrix}_{(N \times N)} \tag{7.19}$$

$$\mathbf{B} = \begin{bmatrix} \frac{\partial^2 x_1}{\partial t \partial w_1} & \frac{\partial^2 x_1}{\partial t \partial w_2} & \cdots & \frac{\partial^2 x_1}{\partial t \partial w_i} & \cdots & \frac{\partial^2 x_1}{\partial t \partial w_M} \\ \frac{\partial^2 x_2}{\partial t \partial w_1} & \frac{\partial^2 x_2}{\partial t \partial w_2} & \cdots & \frac{\partial^2 x_2}{\partial t \partial w_i} & \cdots & \frac{\partial^2 x_2}{\partial t \partial w_M} \\ \vdots & \vdots & \ddots & \vdots & \ddots & \vdots \\ \frac{\partial^2 x_i}{\partial t \partial w_1} & \frac{\partial^2 x_i}{\partial t \partial w_2} & \cdots & \frac{\partial^2 x_i}{\partial t \partial w_i} & \cdots & \frac{\partial^2 x_i}{\partial t \partial w_M} \\ \vdots & \vdots & \ddots & \vdots & \ddots & \vdots \\ \frac{\partial^2 x_N}{\partial t \partial w_1} & \frac{\partial^2 x_N}{\partial t \partial w_2} & \cdots & \frac{\partial^2 x_N}{\partial t \partial w_i} & \cdots & \frac{\partial^2 x_N}{\partial t \partial w_M} \end{bmatrix}_{(N \times M)} \tag{7.20}$$

$$\mathbf{C} = \begin{bmatrix} \frac{\partial y_1}{\partial x_1} & \frac{\partial y_1}{\partial x_2} & \cdots & \frac{\partial y_1}{\partial x_i} & \cdots & \frac{\partial y_1}{\partial x_N} \\ \frac{\partial y_2}{\partial x_1} & \frac{\partial y_2}{\partial x_2} & \cdots & \frac{\partial y_2}{\partial x_i} & \cdots & \frac{\partial y_2}{\partial x_N} \\ \vdots & \vdots & \ddots & \vdots & \ddots & \vdots \\ \frac{\partial y_i}{\partial x_1} & \frac{\partial y_i}{\partial x_2} & \cdots & \frac{\partial y_i}{\partial x_i} & \cdots & \frac{\partial y_i}{\partial x_N} \\ \vdots & \vdots & \ddots & \vdots & \ddots & \vdots \\ \frac{\partial y_Q}{\partial x_1} & \frac{\partial y_Q}{\partial x_2} & \cdots & \frac{\partial y_Q}{\partial x_i} & \cdots & \frac{\partial y_Q}{\partial x_N} \end{bmatrix}_{(Q \times N)} \tag{7.21}$$

$$\mathbf{D} = \begin{bmatrix} \frac{\partial y_1}{\partial w_1} & \frac{\partial y_1}{\partial w_2} & \cdots & \frac{\partial y_1}{\partial w_i} & \cdots & \frac{\partial y_1}{\partial w_M} \\ \frac{\partial y_2}{\partial w_1} & \frac{\partial y_2}{\partial w_2} & \cdots & \frac{\partial y_2}{\partial w_i} & \cdots & \frac{\partial y_2}{\partial w_M} \\ \vdots & \vdots & \ddots & \vdots & \ddots & \vdots \\ \frac{\partial y_i}{\partial w_1} & \frac{\partial y_i}{\partial w_2} & \cdots & \frac{\partial y_i}{\partial w_i} & \cdots & \frac{\partial y_i}{\partial w_M} \\ \vdots & \vdots & \ddots & \vdots & \ddots & \vdots \\ \frac{\partial y_Q}{\partial w_1} & \frac{\partial y_Q}{\partial w_2} & \cdots & \frac{\partial y_Q}{\partial w_i} & \cdots & \frac{\partial y_Q}{\partial w_M} \end{bmatrix}_{(Q \times M)} \tag{7.22}$$

The Jacobians **A**, **B**, **C**, and **D** must be evaluated around the operating point obtained in (7.12). From such a procedure, the linearized matrices $\mathbf{A_{lin}}$, $\mathbf{B_{lin}}$, $\mathbf{C_{lin}}$, and $\mathbf{D_{lin}}$ are obtained. Then, the state-space system of described by (7.17) and (7.18) is rewritten using the linearized matrices to obtain a linear representation of the power converter.

In addition, the linearized system can be transformed into transfer functions using (7.23) to obtain the matrix **H(s)**, which has size $Q \times M$, where each component $H_{i,j}(s)$ is the transfer function between the input j and the output i.

$$\mathbf{H(s)} = \mathbf{C_{lin}} \cdot \left(s \cdot \mathbf{I} - \mathbf{A_{lin}}\right)^{-1} \cdot \mathbf{B_{lin}} + \mathbf{D_{lin}} \tag{7.23}$$

For example, applying this modeling procedure to the averaged differential equations of the boost converter, (7.10) and (7.11), the linear system given in (7.24) is obtained. In (7.18), the output vector $\overline{\mathbf{Y}} = \left[\overline{i_L}\ \overline{v_C}\right]^T$ is defined to analyze both converter states. Then, $\mathbf{C_{lin}}$ and $\mathbf{D_{lin}}$ matrices are given in (7.25). Any other linear combinations of the inputs and states can be used to define the desired output vector.

$$\overbrace{\begin{bmatrix} \frac{\partial \overline{i_L}}{\partial t} \\ \frac{\partial \overline{v_C}}{\partial t} \end{bmatrix}}^{\frac{\partial \overline{\mathbf{X}}}{\partial t}} = \overbrace{\begin{bmatrix} 0 & \frac{-(1-D)}{L} \\ \frac{(1-D)}{C} & \frac{-1}{R \cdot C} \end{bmatrix}}^{\mathbf{A_{lin}}} \cdot \overbrace{\begin{bmatrix} \overline{i_L} \\ \overline{v_C} \end{bmatrix}}^{\overline{\mathbf{X}}} + \overbrace{\begin{bmatrix} \frac{V_C}{L} & \frac{1}{L} \\ \frac{-I_L}{C} & 0 \end{bmatrix}}^{\mathbf{B_{lin}}} \cdot \overbrace{\begin{bmatrix} d \\ v_g \end{bmatrix}}^{\mathbf{U}} \tag{7.24}$$

$$\underbrace{\begin{bmatrix} \overline{i_L} \\ \overline{v_C} \end{bmatrix}}_{\overline{\mathbf{Y}}} = \underbrace{\begin{bmatrix} 1 & 0 \\ 0 & 1 \end{bmatrix}}_{\mathbf{C_{lin}}} \cdot \underbrace{\begin{bmatrix} \overline{i_L} \\ \overline{v_C} \end{bmatrix}}_{\overline{\mathbf{X}}} + \underbrace{\begin{bmatrix} 0 & 0 \\ 0 & 0 \end{bmatrix}}_{\mathbf{D_{lin}}} \cdot \underbrace{\begin{bmatrix} d \\ v_g \end{bmatrix}}_{\mathbf{U}} \tag{7.25}$$

Using (7.23), the four components of $\mathbf{H(s)}$ are obtained:

$$H_{1,1}(s) = \frac{\overline{i_L}(s)}{d(s)} = \frac{V_C + I_L \cdot R - D \cdot I_L \cdot R + C \cdot R \cdot V_C \cdot s}{C \cdot L \cdot R \cdot s^2 + L \cdot s + R \cdot (1 - 2 \cdot D + D^2)} \tag{7.26}$$

$$H_{1,2}(s) = \frac{\overline{i_L}(s)}{v_g(s)} = \frac{1 + R \cdot C \cdot s}{C \cdot L \cdot R \cdot s^2 + L \cdot s + R \cdot (1 - 2 \cdot D + D^2)} \tag{7.27}$$

$$H_{2,1}(s) = \frac{\overline{v_C}(s)}{d(s)} = \frac{R \cdot V_C \cdot \left[(1 - D) - L \cdot I_L \cdot s\right]}{C \cdot L \cdot R \cdot s^2 + L \cdot s + R \cdot (1 - 2 \cdot D + D^2)} \tag{7.28}$$

$$H_{2,2}(s) = \frac{\overline{v_C}(s)}{v_g(s)} = \frac{R \cdot (1 - D)}{C \cdot L \cdot R \cdot s^2 + L \cdot s + R \cdot (1 - 2 \cdot D + D^2)} \tag{7.29}$$

These transfer functions are useful to analyze the dynamic behavior of the boost converter and to design control systems. For example, from $H_{2,1}(s)$ it is deduced that the transfer function between the output voltage and the duty cycle exhibits a non-minimum phase behavior due to a right-half-plane zero [6]. Therefore, classical proportional-integral (PI) and proportional-integral-derivative (PID) controllers to regulate the output voltage are difficult to design, so other structures, such as cascade current-voltage controllers, are required.

7.3 Effects of the Converter in the Power Conversion Chain

This section is focused on the development of a steady-state model of the PV system for long-term simulations and design processes. This study is divided into three subsections: the identification of the steady-state model of the power conversion chain, the analysis and simulation of the PV system using the steady-state model, and the impact of the power converter on the voltage ripple of the PV source.

7.3.1 Steady-state Model of the Power Conversion Chain

The steady-state model of the power conversion chain allows estimation of the energy production of a PV generator including the effect of the power electronics interface.

The model is also useful for testing the performance of MPPT algorithms, including the power losses introduced by the DC/DC converter, but without accounting for a detailed simulation of the power interface, in order to speed up the calculation process. In addition, the model is also useful to calculate both the rating of the power system components and the steady-state condition of each electrical variable.

Figure 7.5 shows the block diagram of the PV system at steady state, where a system of non-linear equations **F** models the PV current in terms of the PV voltage and the array parameters, as shown in Chapters 2, 5, and 6. Similarly, the sets of non-linear equations **G** and **H** model the DC/DC converter, including the effect of the modulator and the converter electrical efficiency η_{pc}, the controllers, and the load voltage V_b, current I_b, or power P_b. So the model of the DC/DC converter can be modified in agreement with the desired input variable: duty cycle ($D = D_{REF}$, open-loop operation), voltage, or current references (V_{REF} and I_{REF}, closed-loop operation), among others.

The equations for the left-hand block – the PV array model – have been obtained in previous chapters, while the equations for the right-hand block are obtained following the procedures reviewed in Section 7.2. To illustrate the calculation of the steady-state model, the MIU depicted in Figure 7.6 is adopted: it consists of a boost converter and a PV module connected to a DC link.

This example considers a boost converter because, in general, grid-connected PV systems must match the high-voltage requirement of inverters with the low-voltage operation of PV generators. In addition, the DC link is modeled by a voltage source because commercial PV inverters regulate the steady-state voltage of the DC link [8, 9]. The MIU also uses a modulator to impose a given duty cycle on the DC/DC converter. Finally, the

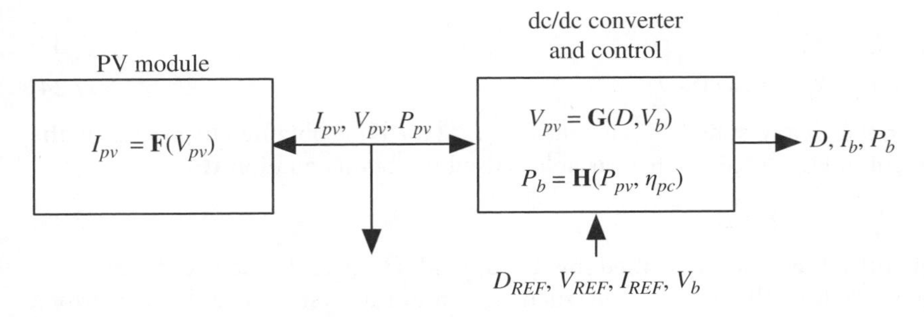

Figure 7.5 Block diagram for the steady-state simulation.

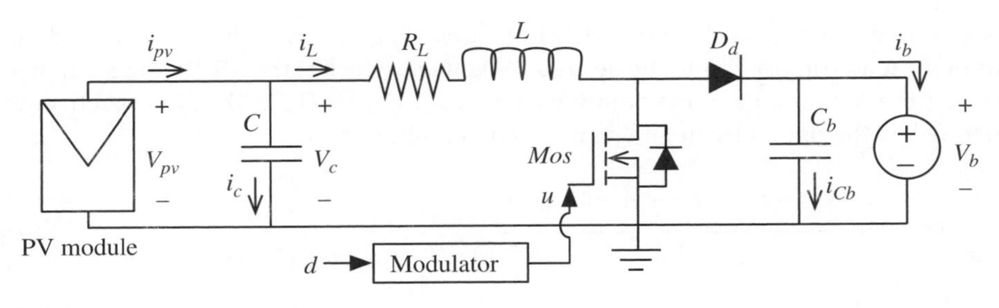

Figure 7.6 MIU based on a boost converter.

resistor R_L models the aggregated ohmic losses introduced by the parasitic resistances of both the inductor and the semiconductors.

This example considers the ideal SDM presented in Section 2.3, which disregards the effects of both the parallel and series resistances of the SDM [10]. Therefore, the current–voltage equation **F** modeling the PV module is given as (7.30), as reported in Section 2.3, where the short-circuit current is $I_{sc} = I_{ph} - A$, the open-circuit voltage is $V_{oc} = (1/B) \cdot \ln\left(I_{ph}/A\right)$, and A and B are parameters extracted from the datasheet or experimental data.

$$i_{pv} = I_{ph} - A \cdot e^{(B \cdot v_{pv})} \tag{7.30}$$

Equations **G** are extracted from the differential equations of the DC/DC converter, where the states are the inductor current i_L and input capacitor voltage $v_C = v_{pv}$. The output capacitor voltage is not considered as a state because its average value is imposed by the inverter. In fact, in the model of Figure 7.6, the voltage of the output capacitor is imposed by the source voltage v_b.

The averaged differential equations modeling the system are:

$$L \cdot \frac{\partial \overline{i_L}}{\partial t} = \overline{v_C} - \overline{i_L} \cdot R_L - v_b \cdot (1 - d) \tag{7.31}$$

$$C \cdot \frac{\partial \overline{v_C}}{\partial t} = i_{pv} - \overline{i_L} \tag{7.32}$$

Then, the steady-state equations **G** are obtained by setting the states' derivatives equal to zero. In such a way, **G** for the MIU in Figure 7.6 is formed by (7.33) and (7.34), where the capital variables represent steady-state values.

$$I_{pv} = I_L, \quad V_{pv} = V_C \tag{7.33}$$

$$V_{pv} - I_{pv} \cdot R_L - V_b \cdot (1 - D) = 0 \tag{7.34}$$

Moreover, the steady-state load current $I_b = I_L \cdot (1 - D)$ is obtained by averaging the diode current in Figure 7.6. Such a variable is used to obtain equation **H**:

$$P_b = I_{pv} \cdot V_b \cdot (1 - D) \tag{7.35}$$

Subsequently, the system formed by **F**, **G**, and **H** must be solved to find the unknown variables. The solution of such a non-linear system can be performed by using analytical or numerical techniques, such as the Newton–Raphson method, or by means of well-known computational applications such as *fsolve* in MATLAB®.

For example, the duty cycle of the MIU in Figure 7.6, given by (7.36), is calculated in terms of the load voltage V_b and the desired PV voltage, which is the MPP voltage. In the same way, the minimum and maximum duty cycles, D_{oc} and D_{sc} in (7.37), respectively, are defined by the open-circuit and short-circuit conditions.

$$D = 1 - \frac{V_{pv} - R_L \cdot \left(I_{ph} - A \cdot e^{(B \cdot V_{pv})}\right)}{V_b} \tag{7.36}$$

$$D_{oc} = 1 - \frac{V_{oc}}{V_b}, \quad D_{sc} = 1 \tag{7.37}$$

Similarly, the steady-state PV voltage imposed by a given duty cycle is analytically calculated using (7.38). Finally, the steady-state inductor current, which has the same value as the steady-state PV current, is calculated from (7.30).

$$V_{pv} = V_b \cdot (1 - D) + R_L \cdot I_{ph} - \frac{W(\theta)}{B} \tag{7.38}$$
$$\theta = R_L \cdot A \cdot B \cdot e^{[B \cdot (V_b \cdot (1-D) + R_L \cdot I_{ph})]}$$

The previous steady-state formulas are useful for calculating the MIU operating point for different load and environmental conditions. Similarly, such equations allow the designer to calculate the rating of the electronic power components and to design their protective circuitry.

Finally, it must be pointed out that using more complex PV models, such as the SDM, the DDM, or the mismatched PV model, would increase the complexity of **F** significantly. Similarly, the adoption of more complex DC/DC converters (such as Sepic, Cuk, or Zeta) increases the complexity of **G** and **H** significantly, so solving analytically the system formed by **F**, **G**, and **H** to obtain explicit expressions, such as given in (7.35)–(7.38), is not possible. Therefore, numerical methods are required to solve the non-linear equation system to find the steady-state values of those PV systems.

7.3.2 Analysis and Simulation using the Steady-state Model

The steady-state model is suitable for analyzing the effect of the DC/DC converter on the power profile delivered to the load. It is assumed that the MIU is the one shown in Figure 7.6, where a BP-585 PV module ($A = 8.95 \times 10^{-7}$ A and $B = 1.406$ V^{-1}) and a parasitic resistance $R_L = 300$ mΩ are adopted.

From the steady-state expressions (7.33)–(7.38), the power curves at both the module terminals (PV power P_{pv}) and the load terminals (load power P_b) are depicted in Figure 7.7 for an irradiance condition $G = 1000$ W/m^2. The efficiency $\eta_{pc} = P_b/P_{pv}$ of the DC/DC converter is also shown. The simulation results show the steady-state effects of the DC/DC converter efficiency on the power delivered to the load. The load power is

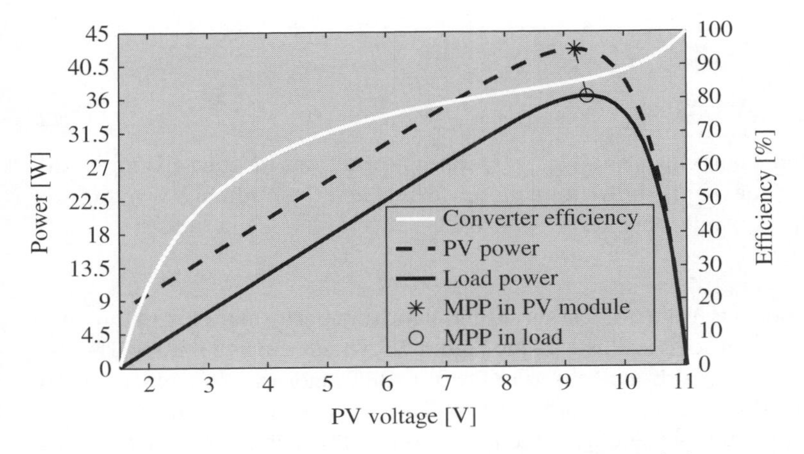

Figure 7.7 Power curves of a MIU based on a BP-585 PV module and a boost converter.

lower than the PV power due to the losses introduced by the DC/DC converter. Moreover, the efficiency of the converter grows at higher PV voltages (lower PV currents) since the ohmic losses decrease.

Another effect concerns the existence of a new optimal operating voltage that maximizes the power delivered to the load. Figure 7.7 shows that the MPP voltage of the PV module does not coincide with the PV voltage ensuring the maximum load power, hence the tracking of the optimal operating condition must be performed by measuring the load power.

Several commercial devices perform the tracking of the optimal conditions – MPPT operation – by maximizing the PV power instead of the load power. This is due to the difficulty of maximizing the load power in grid-connected systems, where the sinusoidal output power could confuse the MPPT algorithm. The operating point with maximum PV power production is calculated by solving (7.39) based on the equation system **F**. For example, the MPP voltage of the MIU analysed in this section is calculated from (7.30), thus obtaining (7.40). Then, the MPP current and power are obtained using (7.30).

$$\frac{dP_{pv}}{dV_{pv}} = 0 \tag{7.39}$$

$$V_{MPP} = \frac{1}{B} \cdot \left[-1 + \mathrm{W}\left(\frac{e^{1} \cdot I_{ph}}{A} \right) \right] \tag{7.40}$$

The same procedure is used to calculate the optimal PV voltage $V_{Max,b}$ that ensures the maximum power delivered to the load. The equations in **F**, **G**, and **H** (Equations (7.30) and (7.33)–(7.38) for the MIU) are used to solve (7.41), which defines the condition for the maximum load power. For example, the value of $V_{Max,b}$ for the MIU is obtained by solving (7.42), which is an implicit equation derived from **F**, **G**, and **H**, where $I_{PV}\left(V_{Max,b}\right)$ represents the PV current at $V_{Max,b}$. Since (7.42) does not have an explicit solution, even involving the Lambert-W function, a numerical method must be used to calculate $V_{Max,b}$. Then, this voltage is used to calculate the maximum load power $P_{Max,b}$ from **H**.

$$\frac{dP_{b}}{dV_{pv}} = 0 \tag{7.41}$$

$$\begin{aligned} &- A \cdot B \cdot V_{Max,b} \cdot e^{\left(B \cdot V_{Max,b}\right)} + \\ &2 \cdot A \cdot B \cdot R_{L} \cdot e^{\left(B \cdot V_{Max,b}\right)} \cdot I_{PV}\left(V_{Max,b}\right) + I_{PV}\left(V_{Max,b}\right) = 0 \end{aligned} \tag{7.42}$$

Figure 7.8 presents both V_{MPP} and $V_{Max,b}$ operating conditions for the MIU under different irradiance values. This shows the difference between the optimal PV voltage for both P_{MPP} and $P_{Max,b}$ conditions.

Simulation of MPPT Algorithms

The steady-state model is useful for testing MPPT algorithm performance with low computational effort. This process is illustrated in Figure 7.9, which shows a flow chart of the testing procedure of a perturb-and-observe (P&O) MPPT algorithm [8], which aims at maximizing the load power instead of the PV power. The process considers the P&O algorithm acting directly on the duty cycle D [8], so (7.38) is used to calculate the PV voltage imposed by the MPPT algorithm.

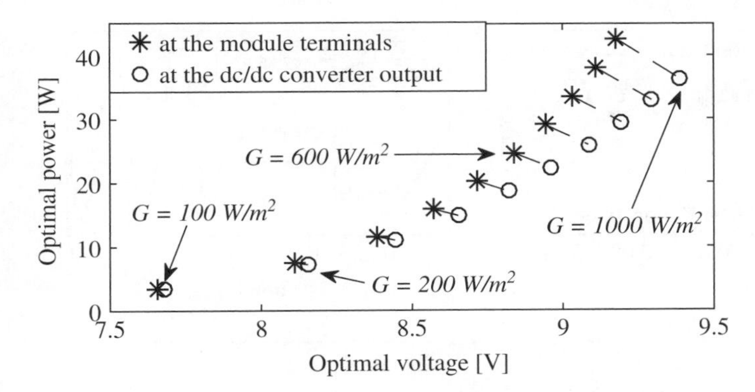

Figure 7.8 Effect of converter efficiency in the optimal PV voltage.

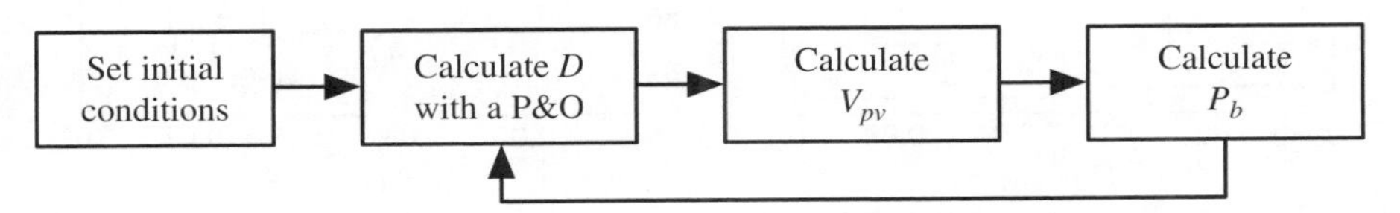

Figure 7.9 Flowchart to test a P&O algorithm using the steady-state model.

Figure 7.10 shows the simulation of the P&O algorithm using the steady-state model. Such a simulation considers a fixed load voltage $V_b = 24\,\text{V}$ for the time interval $[0, 0.06]$ s, where the P&O algorithm converges to stable three-point-profiles [8, 9] for the irradiance conditions $G_1 = 600\,\text{W/m}^2$ at $[0, 0.03]$ s and $G_2 = 1000\,\text{W/m}^2$ at $[0.03, 0.06]$ s. Hence, the P&O algorithm drives the MIU to the optimal PV voltages to maximize the load power, and these are different from the PV MPP voltages. In addition, the simulation considers a sinusoidal perturbation affecting the load voltage V_b with an amplitude that is equal to 15% of the peak voltage (12 V) for the time interval $[0.06, 0.10]$ s. Such a perturbation reproduces a scaled-down effect similar to the bulk voltage oscillation generated by the DC/AC stage operation, in a double-stage inverter connected to the grid [9]. The simulation results show an incorrect operation of the P&O algorithm in such a condition, the PV voltage exhibiting large oscillations, which reduce the load power. Therefore, a classical P&O algorithm is not suitable for optimizing the load power in the presence of load-voltage oscillations. This issue is addressed by introducing an additional controller, the design of which will be discussed in Chapter 8.

Estimation of Power Production

The steady-state model is also useful for estimating the power profile generated by a PV system over long time intervals: days, months, or years. The steady-state model provides several advantages over classical solutions such as simulations carried out in circuit-oriented software or simulations based on PV models considering constant converter efficiencies and voltage sweeps. By using PSIM or PSPICE, a level of accuracy similar to that of the steady-state model can be achieved, but at the price of detailed simulations of the converter dynamics and MPPT controller.

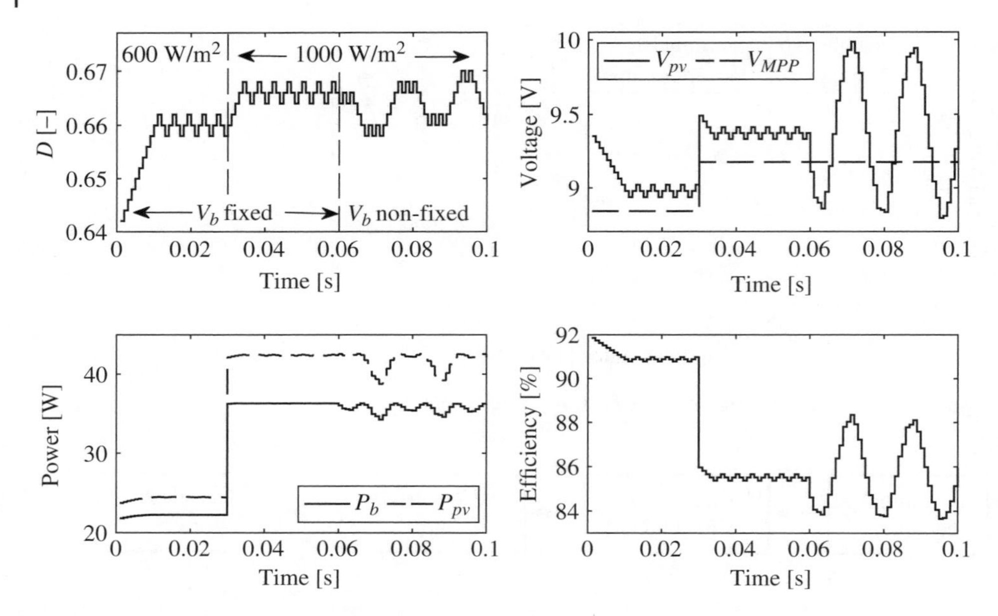

Figure 7.10 Simulation of the P&O algorithm using the steady-state model.

PV models similar to the ones presented in Chapters 3–5 have been adopted by some authors with the objective of determining the effect of mismatching conditions on the PV field power production without accounting for the effect of the MPPT algorithms [11]. In such solutions, the MPP for each irradiance condition is detected by performing voltage sweeps, hence requiring several iterations to calculate the power in each PV voltage value. In addition, the DC/DC converter is usually not considered at all, or it is modeled by an efficiency value that is kept constant for any PV power/voltage/current value. In contrast, the steady-state model can be used to accurately calculate the MPP in each irradiance condition using a single iteration.

Moreover, both classical solutions have two conflicting objectives: providing high accuracy in the MPP calculation and reducing the processing time. In contrast, the steady-state model provides accurate results by solving a single non-linear system.

The MPP power P_{MPP} for each irradiance value is calculated using the state-space model as follows: calculate V_{MPP} from (7.40) and I_{MPP} from (7.30), and then determine $P_{MPP} = V_{MPP} \cdot I_{MPP}$. Similarly, the procedure to calculate the maximum load power $P_{Max,b}$ is the following: obtain $V_{Max,b}$ and $I_{Max,b}$ by solving the equation system formed by (7.30) and (7.42), then obtain $P_{Max,b}$ by solving the equation system formed by (7.35) and (7.36).

To illustrate these procedures, the MIU is simulated for a summer day in southern Italy with the irradiance profile presented in Figure 7.11. The figure presents both P_{MPP} and $P_{Max,b}$ power profiles, where the effect of the DC/DC converter efficiency is observed. In the simulation, the energy generated by the PV module is 6.16% higher than the energy delivered to the load. Hence, if the DC/DC converter is not taken into account, the power overestimation could lead to an incorrect viability analysis. In addition, modeling the DC/DC converter as having a constant efficiency is inaccurate because, as depicted in

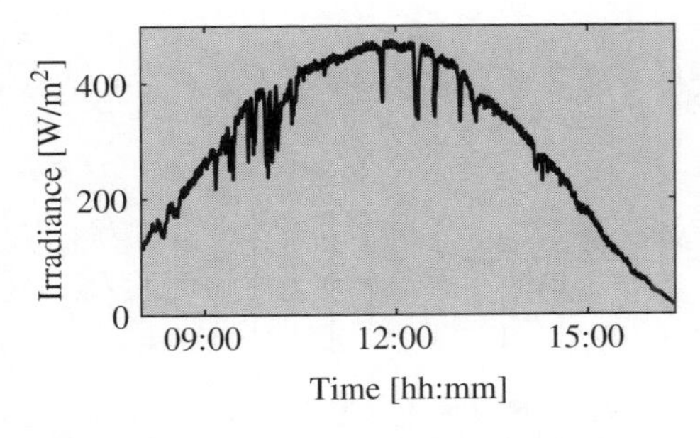

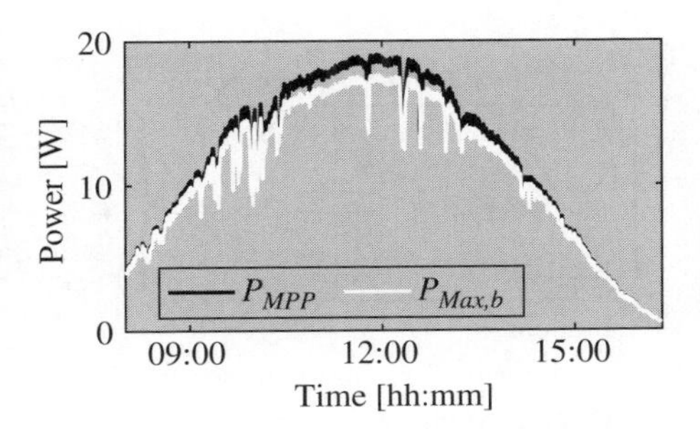

Figure 7.11 Power profile generated by a MIU.

Figure 7.7, the converter efficiency changes with the load power. In particular, the simulation reports a converter efficiency equal to 95.6% at 09:00, 92.5% at 12:00, 97.1% at 15:00, and a maximum efficiency equal to 99.6%.

This simulation also shows that the design of switching converters for PV applications should consider the wide variability of operating conditions, which poses more challenges with respect to the design of DC-DC converters for other applications [12].

Finally, the steady-state model is useful for analyzing the performance of the PV system, including the effect of converter efficiency, but without accounting for the converter dynamics. Similarly, the steady-state model is suitable for testing MPPT algorithms and for calculating the energy production of a PV installation without taking large numbers of iterations, which reduces simulation times compared to classical approaches based on voltage sweeps.

7.3.3 Voltage Ripple at the Generator Terminals

The operation of the DC/DC converter involves a switching operation between different circuital topologies, which generates voltage ripples across the converter capacitors, as described in Section 7.2.1.2. Unfortunately, as illustrated in Figure 7.7, the power produced by a PV module changes significantly, even for small variations of the PV voltage. Therefore, it is important to develop analytical expressions for the design and simulation of voltage ripples in PV systems. Again, the analyses presented in this subsection consider MIUs, such as the boost-based MIU in Figure 7.6, but the procedures can be extended to larger PV arrays by adopting suitable scaled-up PV models.

In general, PV systems require a capacitor between the PV generator and the DC/DC converter, as in Figure 7.6, to filter the current harmonics generated by the switching operation of the converter. The size of the capacitor defines the amplitude of the voltage ripple generated at the PV array terminals.

To obtain expressions for the ripple affecting the PV voltage, the model in Figure 7.12 is designed. In such a model, the DC/DC converter is represented by a current source imposing the waveform generated by the converter. Classical converters are divided into continuous and discontinuous input current topologies [1, 7]. Examples of continuous input current topologies are the boost, Cuk, Sepic, and split-pi converters, while examples of discontinuous input current topologies are the buck, buck–boost, non-inverting buck–boost, and zeta converters.

To illustrate the voltage ripple in PV systems based on both continuous and discontinuous input currents, MIUs based on boost and buck converters are analyzed

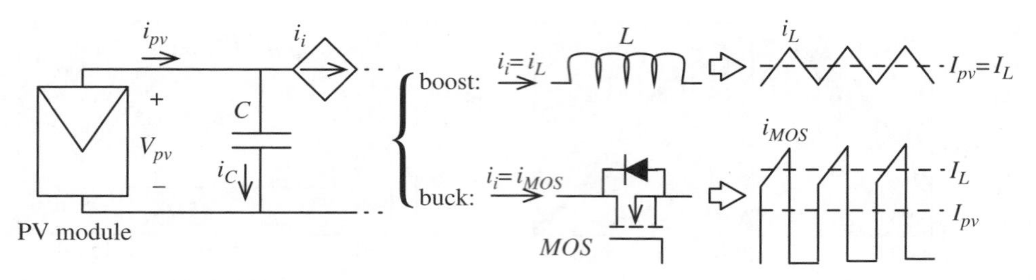

Figure 7.12 Model of the connection between the PV module and the converter.

in Figures 7.6 and 7.26 respectively. In the MIU based on a boost converter, the PV module and the input capacitor interact with the inductor of the converter, so the current source in Figure 7.12 represents the inductor current waveform. In the MIU based on a buck converter, the PV module and the input capacitor interact with the MOSFET of the converter, so the current source in Figure 7.12 represents the MOSFET current waveform. In both cases, the capacitor absorbs the high-frequency components depending on the ripple frequency and PV module impedance.

Since, in general, a PV system includes an MPPT controllers, the PV module operates most of the time around an optimal operating condition. If the MPPT controller measures the PV power, the module operates around the MPP. Instead, if the MPPT controller measures the load power, the module operates around the PV voltage $V_{Max,b}$ that provides the maximum power to the load. Therefore, the PV voltage ripple must be analyzed at those operating conditions. Such analyses must describe the module impedance, which is a resistance, as the ratio of the voltage and current at a given operating point.

Figure 7.13 shows, on the left-hand side, the impedance of the PV module at the MPP as Z_{MPP}, while $Z_{Max,b}$ plots the impedance of the PV module at $V_{Max,b}$ in a MIU based on a boost converter. $V_{Max,b}$ is computed, as shown in Section 7.3.2, by solving (7.42). The figure shows that the module impedance decreases with increasing irradiance values, with impedance values in both V_{MPP} and $V_{Max,b}$ that are close to each other. Moreover, since the module and input capacitance are connected in parallel in the model of Figure 7.12, the current ripple in i_i is distributed across the PV module and the capacitor in inverse proportion to the impedance of each element: small module impedances, in comparison with the capacitor impedance, increase the current ripple reaching the PV module. Therefore, the PV voltage ripple must be analyzed at the highest irradiance condition, where the module impedance assumes its smallest value, this representing the worst-case condition, say $Z_{Max,b} = 2.07\,\Omega$ at $G = 1000\,\text{W/m}^2$ for the BP-585 PV module.

In addition, since the magnitude of the capacitor impedance changes with the frequency f and capacitance C, as in (7.43), the larger component of the voltage ripple corresponds to the switching frequency F_{sw}, since at higher frequencies the capacitor impedance is smaller and the amplitudes of the harmonics are smaller. In the light of the previous considerations and adopting $F_{sw} = 50\,\text{kHz}$, Figure 7.13 presents, on the right, the relationship between the input current ripple Δi_i generated by the converter and the fraction of this ripple Δi_C absorbed by the capacitor, which is given in (7.44). The figure shows that, depending on the capacitance and irradiance, the capacitor is able to absorb a high percentage of the input current ripple. Thus Figure 7.13 shows the performance of the MIU in Figure 7.6 for different input capacitances. In this example,

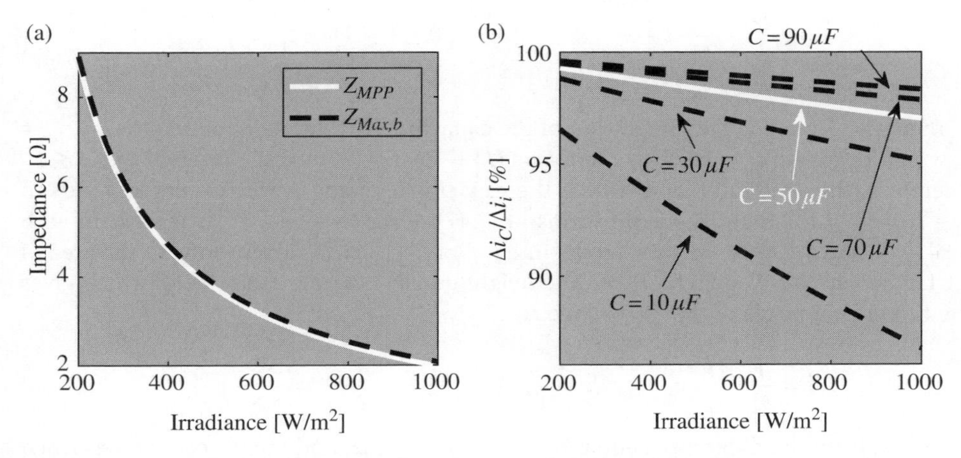

Figure 7.13 Impedance of the PV module at the optimal condition of the MIU in Figure 7.6.

for capacitances higher than $C = 50\ \mu$F, more than 97% of the ripple flows through the capacitor. The simulation also shows that capacitances smaller than 50 μF give a much smaller ripple filtering, while higher capacitances do not improve the ripple filtering significantly.

$$Z_C(f) = \frac{1}{2 \cdot \pi \cdot f \cdot C} \tag{7.43}$$

$$\Delta i_C = \frac{Z_C}{Z_C + Z_{Max,b}} \cdot \Delta i_i \tag{7.44}$$

This analysis shows that an input capacitor with a sufficiently high value ensures that the voltage ripple at the PV terminals is mainly generated by the current that flows through the capacitor.

Analyzing again the model in Figure 7.12, it is noted that the steady-state values of the input current i_i and PV current i_{pv} are equal due to the charge balance in the input capacitor, so the steady-state current in the capacitor is zero [1]. The steady-state value, which is for both the input current I_i and PV current I_{pv}, is calculated from (7.45), so the current ripple at the input capacitor is given by (7.46).

$$I_i = I_{pv} = \frac{1}{T_{sw}} \cdot \int_0^{T_{sw}} i_i\, dt \tag{7.45}$$

$$i_C = I_i - i_i \tag{7.46}$$

Considering the MIU based on a boost converter, its averaged differential equations are given in (7.31) and (7.32) and, following the procedure described in Section 7.2.1.2, the current waveform in the capacitor is given by (7.47), where Δi_L represents the ripple magnitude of the inductor current. Then, the voltage ripple at the capacitor – which is the module terminals – is generated by this current waveform. Hence, the ripple magnitude of the PV voltage $\Delta v_{pv,boost}$ is obtained by integrating the capacitor current when:

- $i_{C,boost} > 0$; when the PV voltage grows
- $i_{C,boost} < 0$; when the PV voltage decreases.

$$i_{C,boost} = \begin{cases} -\frac{\Delta i_L}{D \cdot T_{sw}} t + \frac{\Delta i_L}{2} & 0 \leq t \leq D \cdot T_{sw} \\ \frac{\Delta i_L}{D' \cdot T_{sw}} \left(t - D \cdot T_{sw}\right) - \frac{\Delta i_L}{2} & D \cdot T_{sw} < t \leq T_{sw} \end{cases} \tag{7.47}$$

Then, the limits for the integration of the capacitor current are obtained for $i_{C,boost} = 0$, which results in $t_1 = D \cdot T_{sw}/2$ and $t_2 = (D+1) \cdot T_{sw}/2$, thus the interval $t_1 \leq t \leq t_2$. Therefore, the time interval in which the capacitor current is positive corresponds to $t_2 - t_1 = T_{sw}/2$. Finally, the expressions in (7.47) describe a triangular waveform with peak values $\Delta i_L/2$ and $-\Delta i_L/2$, so the integral of $i_{C,boost} > 0$ corresponds to the area of a triangle with base equal to $T_{sw}/2$ and height equal to $\Delta i_L/2$, as in (7.48), which gives the ripple magnitude of the PV voltage $\Delta v_{pv,boost}$.

$$\Delta v_{pv,boost} = \frac{1}{C} \cdot \int_{t_1}^{t_2} i_{C,boost}\, dt = \frac{1}{C} \cdot \left(\frac{1}{2} \cdot \frac{T_{sw}}{2} \cdot \frac{\Delta i_L}{2} \right) = \frac{\Delta i_L \cdot T_{sw}}{8 \cdot C} \tag{7.48}$$

Finally, from the differential equations (7.31) and (7.32) and the procedure described in Section 7.2.1.2, the ripple magnitude of the inductor current Δi_L for a MIU based on a boost converter is given in (7.49), where V_{pv} represents the steady-state value of the PV voltage.

$$\Delta i_L = \frac{V_{pv} - I_L \cdot R_L}{L} \cdot D \cdot T_{sw} \tag{7.49}$$

The same procedure, applied to a MIU based on a buck converter with averaged differential equations as given in (7.74) and (7.75), results in the capacitor current waveform given in (7.50), where $I_{pv} = D \cdot I_L$ due to the charge balance principle. Similarly, the magnitude of the PV voltage $\Delta v_{pv,buck}$ is obtained by integrating the capacitor current, which within $D \cdot T_{sw} \leq t \leq T_{sw}$ describes a rectangle with base equal to $D' \cdot T_{sw}$ and height equal to $D \cdot I_L$. The ripple magnitude of the PV voltage $\Delta v_{pv,buck}$ is then calculated as in (7.51). Finally, the ripple magnitude of the inductor current Δi_L for a MIU based on a buck converter is given in (7.52), where V_b represents the steady-state value of the load voltage.

$$i_{C,buck} = \begin{cases} -\frac{\Delta i_L}{D \cdot T_{sw}} t - D' \cdot I_L + \frac{\Delta i_L}{2} & 0 \leq t \leq D \cdot T_{sw} \\ D \cdot I_L & D \cdot T_{sw} < t \leq T_{sw} \end{cases} \tag{7.50}$$

$$\begin{aligned} \Delta v_{pv,buck} &= \left| \frac{1}{C} \cdot \int_{D \cdot T_{sw}}^{T_{sw}} i_{C,boost}\, dt \right| = \frac{1}{C} \cdot \left(D' \cdot T_{sw} \cdot D \cdot I_L\right) \\ &= \frac{D \cdot D' \cdot I_L \cdot T_{sw}}{C} \end{aligned} \tag{7.51}$$

$$\Delta i_L = \frac{V_{pv} - V_b - I_L \cdot R_L}{L} \cdot D \cdot T_{sw} \tag{7.52}$$

Figure 7.14a shows the simulation of a MIU based on a boost converter operating around $V_{Max,b}$ with $L = 100\,\mu$H, $G = 1000\,$W/m^2 and $V_b = 24\,$V, where expressions (7.48) and (7.49) accurately predict the ripple magnitudes affecting the PV voltage and inductor current, respectively, for two capacitor values. Similarly, Figure 7.14b shows the simulation of a MIU based on a buck converter operating around $V_{Max,b}$ with $G = 1000\,$W/m^2 and $V_b = 4.2\,$V, but using $L = 42.519\,\mu$H, which is the inductance value that ensures the same current ripple magnitude Δi_L as the MIU based on a

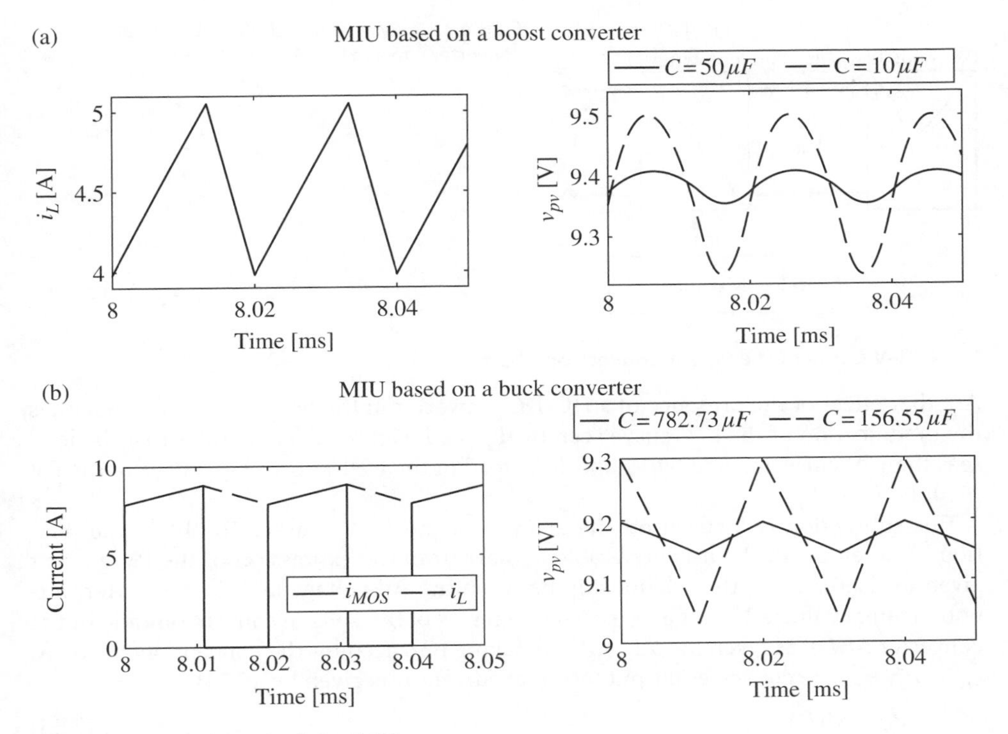

Figure 7.14 PV voltage ripple in a MIU.

boost converter. This choice allows a fair comparison between the two topologies. Similarly, two values for the input capacitor have been selected, so as to provide the same ripple magnitudes in the PV voltages as in the MIU based on the boost converter. With such conditions, (7.51) and (7.52) accurately predict the ripple magnitudes in the PV voltage and inductor current respectively, for both capacitor values.

The simulation results in Figure 7.14 also show the current ripple that must be filtered by the input capacitor in both continuous and discontinuous current topologies: inductor current i_L in the boost case and MOSFET current i_{MOS} in the buck case. These waveforms show that topologies having a discontinuous input current require larger capacitances than topologies having a triangular input current waveform. This condition is also shown by looking at the PV voltage waveforms, where both MIUs provide the same ripple magnitudes affecting the PV voltage, but the MIU based on the buck converter requires a capacitor 15.65 times larger than the one required by the MIU based on a boost converter. In terms of implementation, small capacitances are desirable to avoid the use of electrolytic technologies, which degrade system reliability [13]. Therefore, topologies with continuous input current must be, where possible, adopted over topologies with discontinuous input currents.

The modeling procedure presented in this subsection can be extended to any topology with continuous or discontinuous input currents.

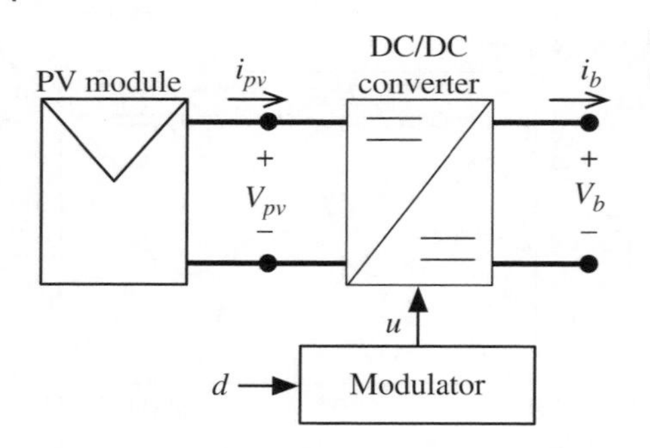

Figure 7.15 PV system including the DC/DC converter effect in the output I-V curve.

7.3.4 I–V Curve of the Power Conversion Chain

Another effect of the presence of a DC/DC converter in the power conversion chain is the modification of the I–V curve seen by the load. Figure 7.15 illustrates that the load sees the I–V curve formed by V_b and I_b instead of the well-known I–V curve of the PV module (I_{pv} vs V_{pv}).

This subsection is focused on the analysis of the I_b–V_b curve. To obtain the relation between I_b and V_b it is preferable to start from the expression of the PV current given in (7.30). Then, the relationship between the PV voltage and the converter output terminal voltage V_b can be expressed in terms of the voltage conversion ratio of the converter $M(D)$, as given in (7.53). Then, taking into account the concept of efficiency, $\eta_{pc} = P_b/P_{pv}$, the converter output terminal current I_b is given by (7.54).

$$V_b = M(D) \cdot V_{pv} \tag{7.53}$$

$$I_b = \frac{P_b}{V_b} = \frac{\eta_{pc} \cdot P_{pv}}{V_b} = \frac{\eta_{pc} \cdot V_{pv} \cdot I_{pv}}{V_b} = I_{pv} \cdot \frac{\eta_{pc}}{M(D)} \tag{7.54}$$

Finally, the current-vs-voltage relation at the converter output terminals is given by (7.55), which shows that the I–V curve at the converter output terminals (I_b vs V_b) changes depending on the duty cycle imposed on the converter.

$$I_b = \frac{\eta_{pc}}{M(D)} \cdot \left[I_{ph} - A \cdot e^{\left(B \cdot \frac{V_b}{M(D)}\right)} \right] \tag{7.55}$$

To illustrate the duty-cycle effect, the same electrical parameters adopted in the previous subsections are used with a constant irradiance of 1000 W/m^2, with the same boost DC/DC converter presented in Figure 7.6, except for the regulated bulk voltage. By exploiting the analyses presented in Section 7.3.1, which correlate the voltages and currents at both the input and output of the DC/DC converter, the I_b-vs-V_b curve is constructed as follows: the output voltage V_b is swept, then the corresponding PV voltage is calculated using (7.38). This depends on the converter duty cycle D, so a particular duty-cycle value must be fixed to construct the curve. Then, the PV current is calculated using (7.30), or any other PV model that might be adopted. Finally, taking into account that at steady state $I_{pv} = I_L$ and $I_b = I_L \cdot (1 - D)$, as given in (7.33), the I_b current is calculated. Figure 7.16 presents the simulation results, where the effect of the DC/DC converter is observed: the white trace represents the I–V curve of the PV module, while the black traces are the I_b–V_b curves at different duty cycles.

The translation of the module PV curve provided by the DC/DC converter operation means that the MPP is translated to a new voltage value V_b depending on the duty-cycle

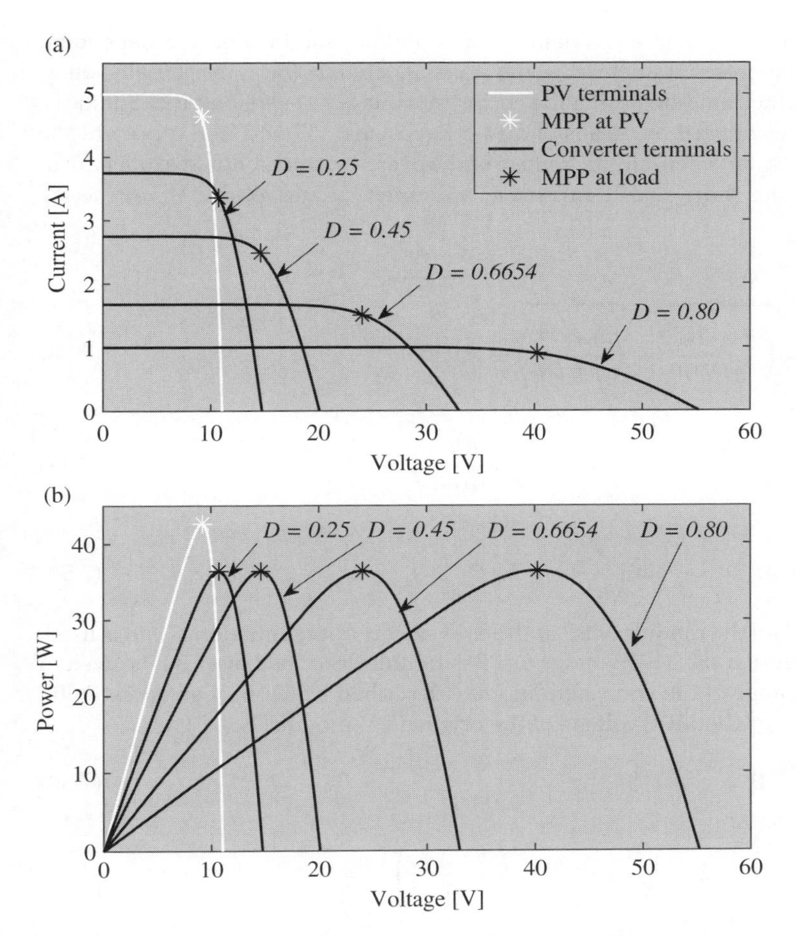

Figure 7.16 Effect of the DC/DC converter on the output I–V curve.

value. For example, the current-vs-voltage characteristic of the power conversion chain can be moved from one of the black traces of Figure 7.16 up to another one to keep the load operating at the MPP condition simply by appropriately changing the duty-cycle value. This could be especially useful for mitigating load perturbations that degrade the performance of classical MPPT algorithms, such as the voltage oscillations caused by grid connection, as illustrated in Figure 7.10. This concept has been used to calculate the duty cycle compensation required to mitigate the 100-Hz voltage oscillations present at the bulk capacitor in double-stage grid-connected PV systems [14].

Figure 7.16 also shows the drop of the power provided to the load due to the DC/DC converter efficiency. Figure 7.16 shows the DC/DC converter output I–V curve at $D = 0.6654$, which is the duty-cycle value that gives the maximum power transfer to the load in the analysis presented in Figure 7.7 (Section 7.3.2) for a regulated bulk voltage $V_b = 24$ V, thus giving the same maximum power at the converter output terminals.

A different approach to (7.55) requires constructing an equation of an equivalent PV module formed by the equations of both the PV module and the DC/DC converter. In this case, (7.56) gives the current-vs-voltage relationship at the DC/DC converter output as if the whole system were an equivalent PV module that can be adjusted depending on the load requirements. Parameters I_{ph} and A are affected by the converter efficiency, but they can be further modified using the converter voltage conversion ratio. Similarly, parameter B can be adjusted by means of $M(D)$. Equations 7.57 and 7.58 represent the short-circuit current and open-circuit voltage of the overall system, both in explicit form and in relation to the short-circuit current I_{sc} and open-circuit voltage V_{oc} of the PV module.

$$I_b = \overbrace{\left(\frac{\eta_{pc} \cdot I_{ph}}{M(D)}\right)}^{\text{Equivalent } I_{ph}} - \overbrace{\left(\frac{\eta_{pc} \cdot A}{M(D)}\right)}^{\text{Equivalent } A} \cdot e^{\overbrace{\left(\frac{B}{M(D)}\right)}^{\text{Equivalent } B} \cdot V_b} \tag{7.56}$$

$$\overbrace{I_b\big|_{V_b=0}}^{\text{Equivalent } I_{sc}} = \left(\frac{\eta_{pc}}{M(D)}\right) \cdot (I_{ph} - A) = \left(\frac{\eta_{pc}}{M(D)}\right) \cdot I_{sc} \tag{7.57}$$

$$\overbrace{V_b\big|_{I_b=0}}^{\text{Equivalent } V_{oc}} = M(D) \cdot \frac{1}{B} \cdot \ln\left(\frac{I_{ph}}{A}\right) = M(D) \cdot V_{oc} \tag{7.58}$$

In order to translate the module MPP at the converter output, Equation (7.59) is used. Taking into account that the MPP voltage of a PV module described by (7.30) is given in (7.40), the MPP voltage of the equivalent module described by (7.56) is given in (7.60), where V_{MPP} stands for the MPP voltage of the original PV module.

$$\frac{dP_b}{dV_b} = 0 \tag{7.59}$$

$$\overbrace{V_{MPP,eq}}^{\text{Equivalent } V_{MPP}} = \begin{cases} M(D) \cdot \frac{1}{B} \cdot \left[-1 + \mathrm{W}\left(\frac{e^1 \cdot I_{ph}}{A}\right)\right] \\ M(D) \cdot V_{MPP} \end{cases} \tag{7.60}$$

To illustrate the concept of the equivalent PV module, Figure 7.17a presents the current-vs-voltage curve of the equivalent PV modules (black traces) for different $M(D)$ values and by assuming the adoption of different DC/DC converter topologies:

- boost conditions: $M(D) = 1.5$ and $M(D) = 2$
- buck conditions: $M(D) = 0.75$ and $M(D) = 0.5$.

Simulations results have been obtained by considering $\eta_{pc} = 1$ for simplicity. Figure 7.17a also presents the I–V curve of the original PV module as a white trace.

These simulations show the ability of the overall system, including the PV source and the DC/DC converter, to adjust the MPP voltage and current to the values required by the load. It is worth noting that DC/DC converters have limitations in terms of voltage and current ratings that must be respected to avoid component damage. In this regard, Figure 7.17b shows the current-vs-voltage curves accounting for the limitations imposed by current and voltage ratings: 8 A and 16 V, respectively. Power ratings are not considered because a properly designed DC/DC converter must be able to process the MPP power of the associated PV module.

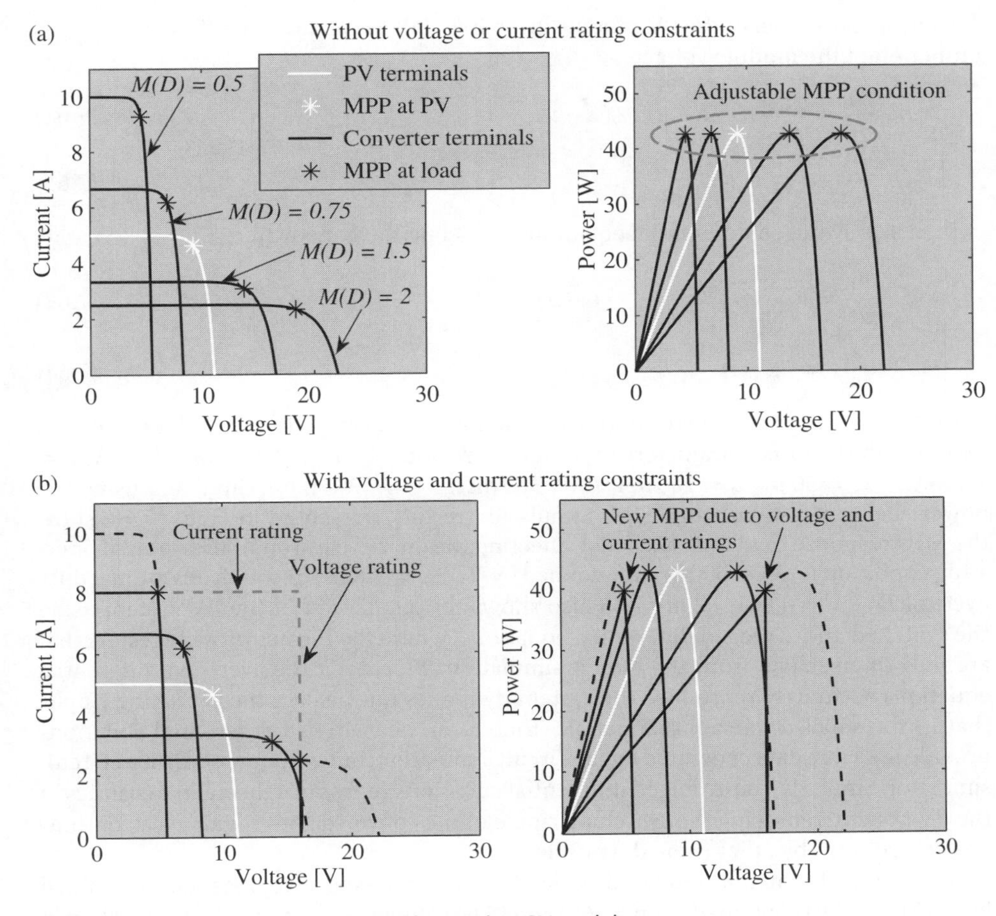

Figure 7.17 Current-vs-voltage curves of equivalent PV modules.

The simulations results in Figure 7.17b reveal that new MPP conditions for $M(D) = 0.5$ and $M(D) = 2$ appear due to the current and voltage ratings, respectively. This condition must be analyzed during the converter design process to avoid undesirable power drops or component damage.

Finally, the modeling approach shown in this chapter is useful for estimating, realistically, the energy production of a PV installation, as reported in Section 3.4.1.

7.4 Modelling the Dynamics of the Power Conversion Chain

This section is focused on modeling the dynamic operation of the PV system to provide formulas that are oriented to control design and stability analysis. This study is illustrated using a MIU based on a boost converter – the circuit in Figure 7.6 – but the analyses can be extended to any DC/DC converter.

Using the procedure described in Section 7.2, the switched equations of the MIU in Figure 7.6 are obtained, as in (7.61) and (7.62), where $v_C = v_{pv}$. These equations also use the non-linear model of the PV module given in (7.30) to calculate the PV current i_{pv} as a function of the module voltage.

$$L \cdot \frac{\partial i_L}{\partial t} = v_{pv} - i_L \cdot R_L - v_b \cdot (1 - u) \tag{7.61}$$

$$C \cdot \frac{\partial v_{pv}}{\partial t} = i_{pv} - i_L \tag{7.62}$$

Then, the averaged differential equations modeling the system are:

$$L \cdot \frac{\partial \overline{i_L}}{\partial t} = \overline{v_{pv}} - \overline{i_L} \cdot R_L - v_b \cdot (1 - d) \tag{7.63}$$

$$C \cdot \frac{\partial \overline{v_{pv}}}{\partial t} = i_{pv} - \overline{i_L} \tag{7.64}$$

To illustrate the accuracy of both the switched and averaged differential equations, a MIU with the same parameters used in the previous section ($A = 8.95 \times 10^{-7}$ A, $B = 1.406\ \mathrm{V}^{-1}$, $C = 50\ \mu$F, $L = 100\ \mu$H, $R_L = 300$ mΩ, $F_{sw} = 50$ kHz) is simulated using the power electronics simulator PSIM. Simulation results, presented in Figure 7.18, show the MIU response to a 2% step change affecting the duty cycle around the optimal operating condition for $G = 500\ \mathrm{W/m^2}$, defined by $V_{Max,b} = 8.825$ V; so with an optimal duty cycle of $D_{Max,b} = 0.661$. Figure 7.18 also shows the simulation of the PV system using the switched and averaged differential equations, where the former provide results that are indistinguishable from the circuit simulation. Moreover, the averaged differential equations accurately reproduce the system dynamics but neglect the switching ripple; that is, the small-ripple approximation. Therefore, the switched differential equations provide the same data generated by the circuit simulation, but without requiring a circuit simulator. Similarly, the averaged differential equations reproduce the main dynamics of the PV system, removing the switching ripple of the converter, but requiring a continuous control variable, that is the duty cycle.

The averaged dynamic model, described in the state-space representation defined in (7.17)–(7.18), is obtained from (7.63) and (7.64) following the procedure described in Section 7.2.2. The averaged state and input vectors of the MIU in Figure 7.6 are defined as $\overline{\mathbf{X}} = \left[\overline{i_L}\ \overline{v_{pv}}\right]^T$ and $\mathbf{U} = \left[d\ v_b\ i_{ph}\right]^T$, respectively, and the state-space system is given in

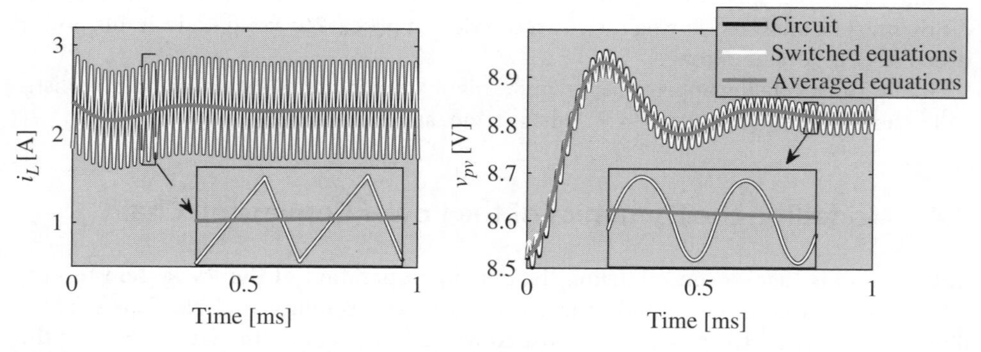

Figure 7.18 Comparison between circuit simulation, switched equations and averaged equations for the MIU in Figure 7.6.

(7.65)–(7.66). This system provides information on both PV voltage and inductor current values. To the aim of designing a classical controller to regulate the PV voltage only, (7.66) has to be modified into $[\overline{v_{pv}}] = [0\ 1]^T \cdot \overline{\mathbf{X}}$.

$$\underbrace{\begin{bmatrix} \frac{\partial \overline{i_L}}{\partial t} \\ \frac{\partial \overline{v_{pv}}}{\partial t} \end{bmatrix}}_{\frac{\partial \overline{\mathbf{X}}}{\partial t}} = \underbrace{\begin{bmatrix} \frac{-R_L}{L} & \frac{1}{L} \\ \frac{-1}{C} & \frac{1}{C}\cdot\frac{\partial i_{pv}}{\partial v_{pv}} \end{bmatrix}}_{\mathbf{A}} \cdot \underbrace{\begin{bmatrix} \overline{i_L} \\ \overline{v_{pv}} \end{bmatrix}}_{\overline{\mathbf{X}}} + \underbrace{\begin{bmatrix} \frac{v_b}{L} & \frac{-(1-d)}{L} & 0 \\ 0 & 0 & \frac{1}{C} \end{bmatrix}}_{\mathbf{B}} \cdot \underbrace{\begin{bmatrix} d \\ v_b \\ i_{ph} \end{bmatrix}}_{\mathbf{U}} \tag{7.65}$$

$$\underbrace{\begin{bmatrix} \overline{i_L} \\ \overline{v_{pv}} \end{bmatrix}}_{\overline{\mathbf{Y}}} = \underbrace{\begin{bmatrix} 1 & 0 \\ 0 & 1 \end{bmatrix}}_{\mathbf{C}} \cdot \underbrace{\begin{bmatrix} \overline{i_L} \\ \overline{v_{pv}} \end{bmatrix}}_{\overline{\mathbf{X}}} + \underbrace{\begin{bmatrix} 0 & 0 & 0 \\ 0 & 0 & 0 \end{bmatrix}}_{\mathbf{D}} \cdot \underbrace{\begin{bmatrix} d \\ v_b \\ i_{ph} \end{bmatrix}}_{\mathbf{U}} \tag{7.66}$$

Then, to obtain a linear state-space model to analyze the small-signal dynamics of the MIU, the system including (7.65)–(7.66) must be linearized around the operating point, as described in Section 7.2.2.

Equation 7.65 reveals that matrix **A** depends on the converter parasitic resistance R_L, on the values of the inductor L and of the input capacitor C, and on the derivative of the PV current with respect to the PV voltage $\partial i_{pv}/\partial v_{pv}$. Therefore, considering constant the electrical parameters of the converter, R_L, L, and C, the natural dynamics of the PV system are affected mainly by the term $\partial i_{pv}/\partial v_{pv} < 0$, which takes into account the effect of the irradiance and temperature on the system dynamics. Considering the non-linear model of the PV module given in (7.30), the term $\partial i_{pv}/\partial v_{pv} = -A \cdot B \cdot e^{(B \cdot v_{pv})}$. The derivative can be calculated also if more detailed models of the PV source are used.

Figure 7.19 presents the values of $\partial i_{pv}/\partial v_{pv}$ for the BP-585 PV module in different irradiance and voltage conditions, with the operating conditions at the optimal voltages V_{MPP} and $V_{Max,b}$, that are duly highlighted. This figure shows the increment in the magnitude of $\partial i_{pv}/\partial v_{pv}$, namely $|\partial i_{pv}/\partial v_{pv}|$, caused by increments in the module voltage at all the irradiance conditions.

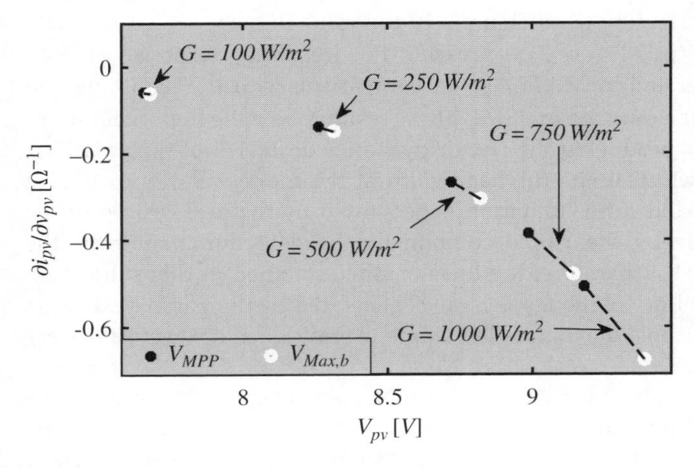

Figure 7.19 Values of $\partial i_{pv}/\partial v_{pv}$ for a BP-585 PV module.

Taking into account that, in general, PV systems include MPPT controllers, the PV module operates most of the time around the optimal operating condition: the MPP if the MPPT optimizes the PV power, or the operating point defined by the PV voltage $V_{Max,b}$ if the MPPT optimizes the load power. Hence Figure 7.19 shows the values of $\partial i_{pv}/\partial v_{pv}$ under both the V_{MPP} and $V_{Max,b}$ conditions. Such simulation results show the small value of $\partial i_{pv}/\partial v_{pv}$ for the optimal conditions of a MIU based on a BP-585 PV module: $[-0.68, -0.06]\ \Omega^{-1}$. It is also evident that the value of $\partial i_{pv}/\partial v_{pv}$ at V_{MPP} or $V_{Max,b}$ changes significantly depending on the irradiance. For instance, it changes by 1133% when irradiance steps up from 100 W/m^2 to 1000 W/m^2. Therefore, very different dynamic behaviors of the MIU are expected for different irradiance conditions.

On the other hand, from (7.65) it is noted that matrix **B** depends on the converter inductor L and input capacitor C values, on its duty cycle d, and on the load voltage v_b. Therefore, considering constant the electrical parameters of the converter, the dynamic behavior of the PV system is mainly affected by the converter operating point.

The dynamics of the states in the PV system in (7.65)–(7.66), with respect to the command introduced in the duty cycle, are modeled by the transfer functions given in (7.67) and (7.68).

$$G_{v_{pv}/d}(s) = \frac{\overline{v_{pv}}(s)}{d(s)} = \frac{\frac{-v_b}{L\cdot C}}{s^2 + \left(\frac{R_L}{L} - \frac{1}{C}\cdot\frac{\partial i_{pv}}{\partial v_{pv}}\right)\cdot s + \left(\frac{1-R_L\cdot\frac{\partial i_{pv}}{\partial v_{pv}}}{L\cdot C}\right)} \tag{7.67}$$

$$G_{i_L/d}(s) = \frac{\overline{i_L}(s)}{d(s)} = \frac{\frac{v_b}{L}\cdot\left(s - \frac{1}{C}\cdot\frac{\partial i_{pv}}{\partial v_{pv}}\right)}{s^2 + \left(\frac{R_L}{L} - \frac{1}{C}\cdot\frac{\partial i_{pv}}{\partial v_{pv}}\right)\cdot s + \left(\frac{1-R_L\cdot\frac{\partial i_{pv}}{\partial v_{pv}}}{L\cdot C}\right)} \tag{7.68}$$

To illustrate the accuracy of the model, the parameters previously adopted to simulate the MIU are used to linearize the system formed by (7.67) and (7.68) around the optimal operating condition $V_{Max,b} = 8.825$ V; thus duty cycle $D_{Max,b} = 0.661$, for irradiance and load-voltage conditions $G = 500$ W/m^2 and $v_b = 24$ V, respectively. Such a linearization requires extracting from Figure 7.19 the value $\partial i_{pv}/\partial v_{pv} = -0.3064\ \Omega^{-1}$ at $V_{Max,b}$ and $G = 500$ W/m^2. Figure 7.20 presents the frequency response of both $G_{v_{pv}/d}(s)$ and $G_{i_L/d}(s)$ generated in MATLAB® in comparison with the data obtained from the non-linear circuit simulated in PSIM. These results show the high accuracy of the transfer functions in reproducing the circuit dynamics up to 1/5 of the switching frequency $F_{sw} = 50$ kHz, which limits the bandwidth of the model validity to 10 kHz in this example. Such a bandwidth limitation is generated by the averaging process: near the switching frequency the ripple components become dominant over the low-frequency dynamics [1]. Therefore, the linear model obtained in this subsection is useful for analyzing the low-frequency dynamics generated by the passive elements but not for analyzing the ripple dynamics. The ripple magnitude and waveforms were analyzed in the Section 7.3.3.

The PV voltage dynamics affect the design of MPPT controllers [8, 9]; for example, the period of a P&O algorithm must be longer than the settling time of the PV voltage, thus of PV power, to avoid unstable behavior in the system. Similarly, the performance of controllers designed using linear models, such as the one given in (7.67), change

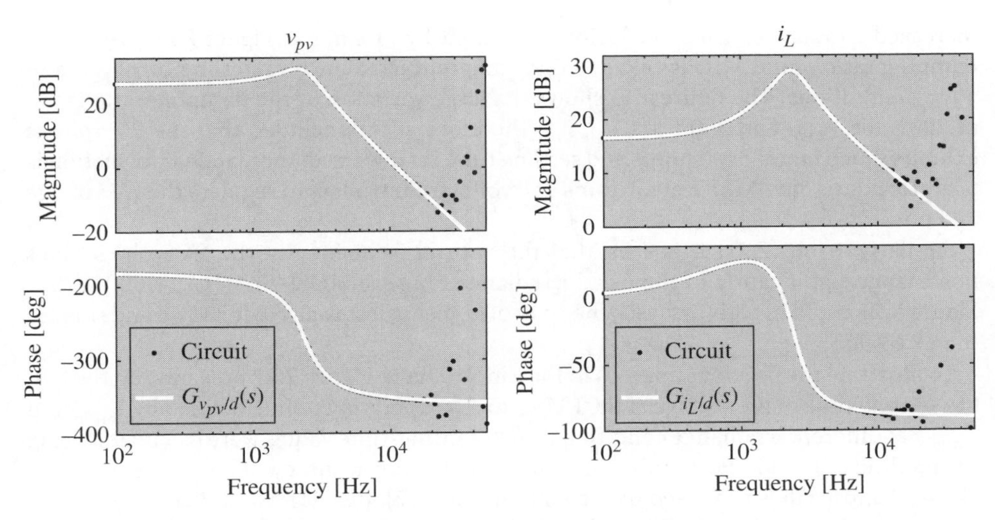

Figure 7.20 Frequency response of circuit and model representing the MIU in Figure 7.6.

when the system moves far from the operating point around which the model was linearized. Therefore, the changes in $G_{v_{pv}/d}(s)$, (Equation 7.67), give information concerning the changes in the dynamics of the PV voltage. In this way, $G_{v_{pv}/d}(s)$ can be expressed as the canonical second-order transfer function given in (7.69), where ρ and ω_n represent the damping ratio and natural frequency of the system, respectively. Therefore, the PV voltage exhibits the damping ratio and natural frequency given in (7.70) and (7.71), respectively.

$$G_{2nd}(s) = A_0 \cdot \frac{\omega_n^2}{s^2 + 2 \cdot \rho \cdot \omega_n \cdot s + \omega_n^2} \tag{7.69}$$

$$\rho = \frac{1}{2} \cdot \frac{R_L \cdot C - L \cdot \frac{\partial i_{pv}}{\partial v_{pv}}}{\sqrt{L \cdot C \cdot \left(1 - R_L \cdot \frac{\partial i_{pv}}{\partial v_{pv}}\right)}} \tag{7.70}$$

$$\omega_n = \sqrt{\frac{1}{L \cdot C} \cdot \left(1 - R_L \cdot \frac{\partial i_{pv}}{\partial v_{pv}}\right)} \tag{7.71}$$

Equation 7.70 reveals that the damping ratio of the PV system increases when L, C, and R_L increase. But changes in $\partial i_{pv}/\partial v_{pv}$ mainly affect ρ. Moreover, Figure 7.19 shows that the magnitude of $\partial i_{pv}/\partial v_{pv}$, at both optimal conditions V_{MPP} and $V_{Max,b}$, is smaller than $0.68\,\Omega^{-1}$, while the parasitic resistance of DC/DC converters is commonly in the milliohm range. Therefore, the term $\left(1 - R_L \cdot \frac{\partial i_{pv}}{\partial v_{pv}}\right)$ in (7.70) is almost constant in the usual irradiance range of $100\,\text{W/m}^2 \leq G \leq 1000\,\text{W/m}^2$. This analysis holds for other PV modules with the same or higher numbers of cells in series, which is common for commercial installations, where the modules are formed by 18, 36, or even more series-connected cells.

Equation 7.70 suggests that an increase of the magnitude of $\partial i_{pv}/\partial v_{pv}$ leads to an increase of ρ, which means that the damping ratio of the PV voltage increases with increased irradiance. This conclusion is verified by looking at Figure 7.21, where the damping ratio of the MIU is calculated for different irradiance values in both V_{MPP} and $V_{Max,b}$ conditions. The figure also shows the large variation of the damping ratio (span of 290% for V_{MPP} and 350% for $V_{Max,b}$). Therefore, it is concluded that the PV voltage exhibits much larger overshoots and settling times at low irradiance. Such information is useful for designing MPPT algorithms and voltage controllers to regulate the PV voltage and power.

Similarly, from (7.71) it is seen that the natural frequency of the PV voltage does not change significantly in the usual irradiance range of $100\,W/m^2 \leq G \leq 1000\,W/m^2$. Figure 7.21 confirms this analysis, the span of ω_n being less than 6% for V_{MPP} and smaller than 8.6% for $V_{Max,b}$.

To illustrate the previous analysis in the time domain, Figure 7.22 presents, on the left, the settling time of the voltage at the PV v_{pv} for the operating points defined by V_{MPP} and $V_{Max,b}$ at different irradiance conditions. Such settling-time values were calculated using the method of Ramos-Paja et al. [15], which provides an accuracy that is higher than the classical approaches proposed by Ogata [16], Kuo [17], Bert [18], and Piche [19].

These settling-time values are useful to design data-logging systems and MPPT algorithms, for example to define the minimum sampling rate for data acquisition or the minimum P&O period to avoid unstable behavior of the PV system. To illustrate the

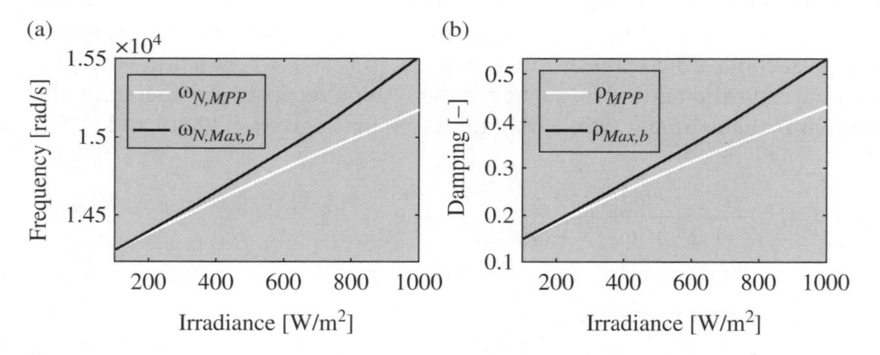

Figure 7.21 Natural frequency and damping of the PV voltage in the MIU in Figure 7.6.

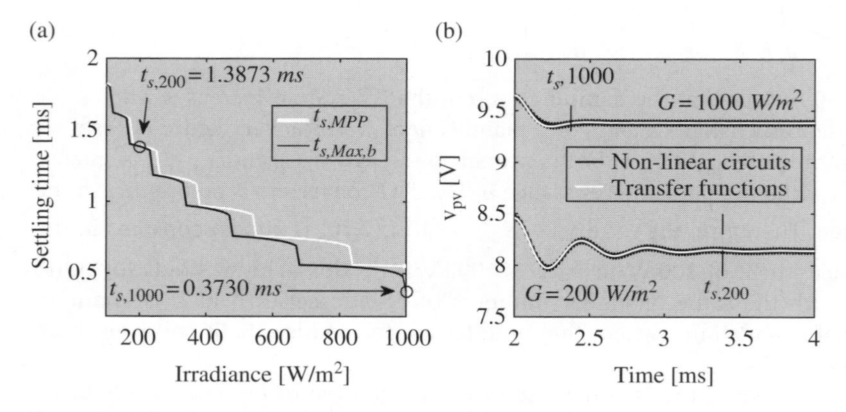

Figure 7.22 Settling time and time response of the PV voltage.

settling time and damping variation caused by the irradiance conditions, Figure 7.22 presents, on the right-hand side, the time response of the PV voltage at 1000 W/m^2 and 200 W/m^2. These simulation results were carried out using both the non-linear switched equations given in (7.61) and (7.62) and the linearized transfer function (7.67). The accuracy of the linear model and linearization process is evident.

Due to the accuracy of the linear model, it is possible to predict the changes in the PV voltage behavior by analyzing the control-to-output transfer function (Equation 7.67). In this way, changes in the operating conditions imposed by external sources such as irradiance and load conditions can be analyzed, as shown in Figure 7.23. In particular, Figure 7.23a presents the frequency response of the PV voltage for different irradiance conditions, where the damping ratio and bandwidth variation (affecting also the settling time) previously described are evident. This "Bode diagram" is useful for designing control systems in the frequency domain, and also to perform robustness analyses to ensure stability in a given irradiance range.

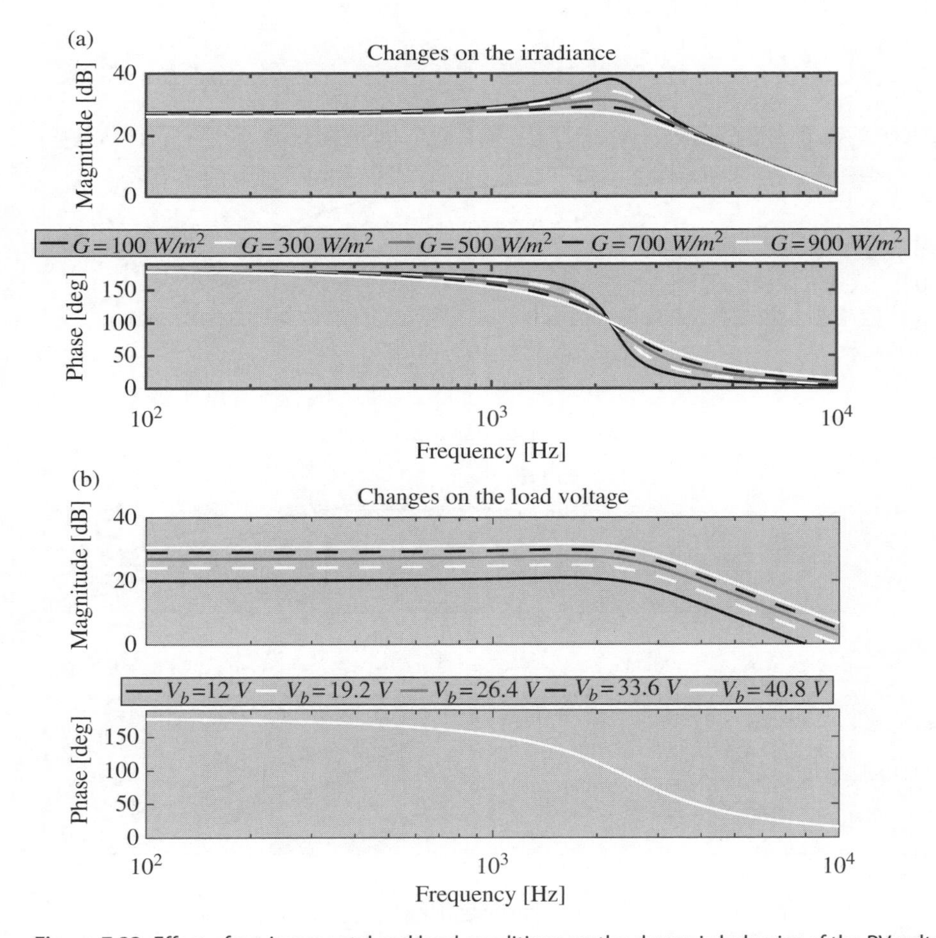

Figure 7.23 Effect of environmental and load conditions on the dynamic behavior of the PV voltage.

Similarly, Figure 7.23b shows the Bode diagrams of the PV voltage for different load–voltage v_b values, where the DC gain directly depends on the load voltage. Instead, the damping ratio and the natural frequency are not affected by variations in v_b. These results show that v_b variations affect the gain margins of PV voltage controllers. Therefore, the behavior of PV voltage controllers must be evaluated in the whole range of v_b to ensure system stability. Such a robustness analysis could be supported by Bode diagrams similar to the one shown in Figure 7.23b.

The linear model is also useful for analyzing the effect of parametric drifts affecting the converter. In this way, the two parts of Figure 7.24 show the frequency responses of the PV voltage for different inductor and capacitor values, respectively. The decrements

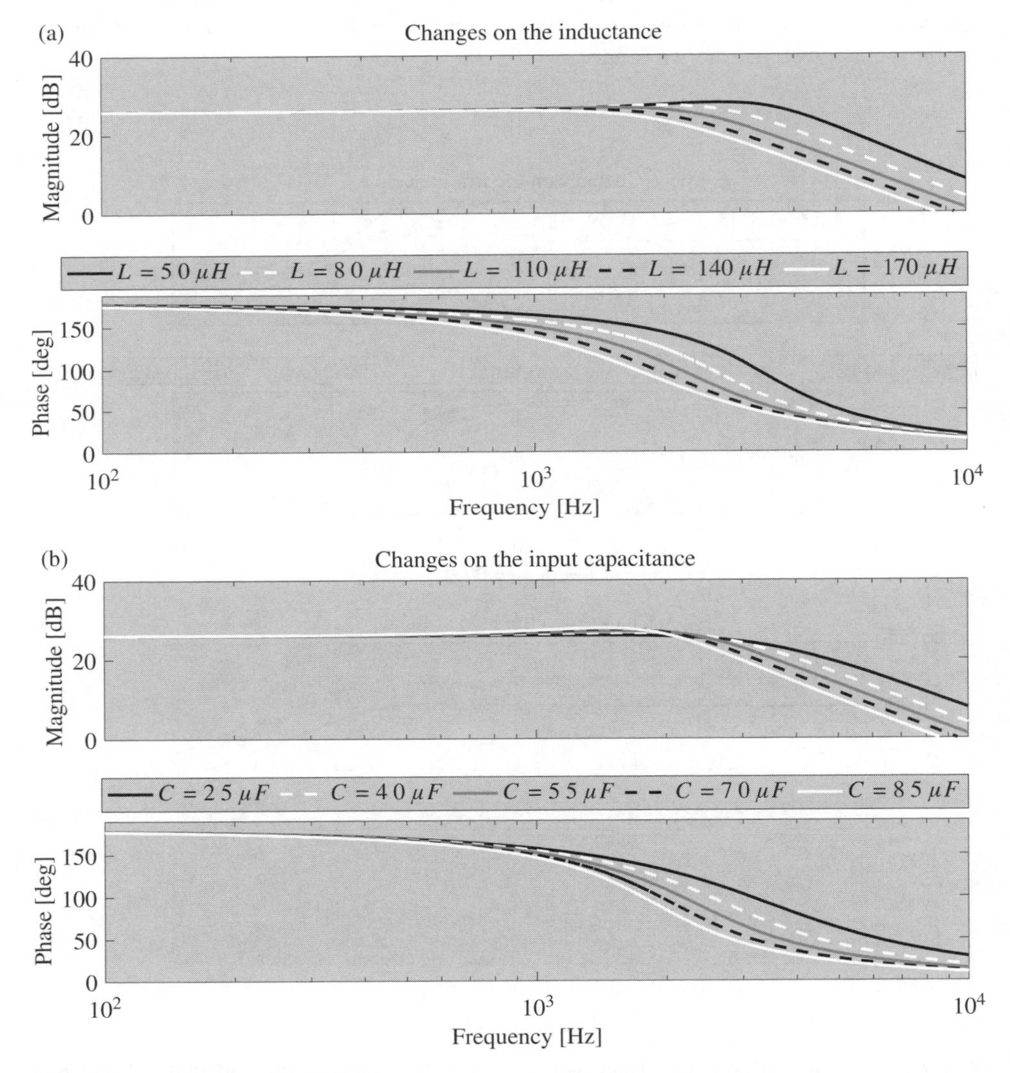

Figure 7.24 Effect on the dynamic behavior of the PV voltage of variations in the converter components (*L* and *C*).

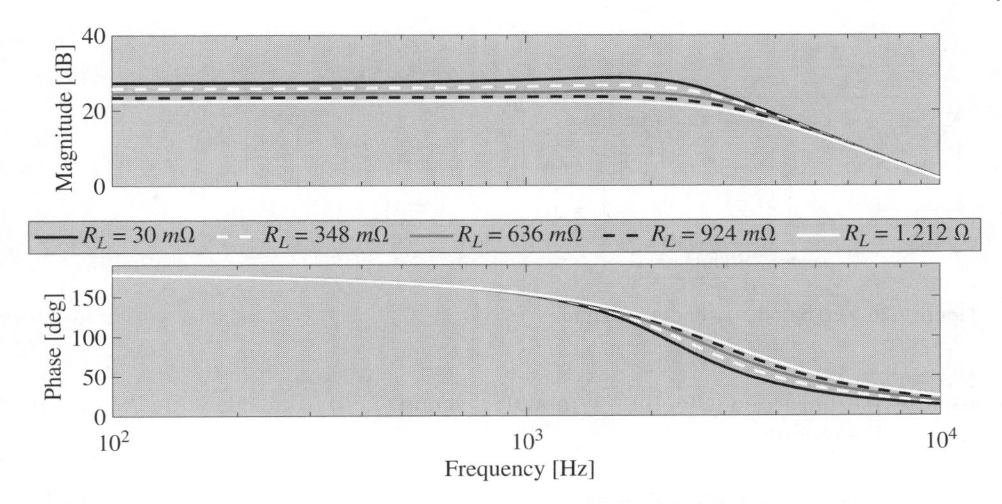

Figure 7.25 Effect on the dynamic behavior of the PV voltage of variations in the converter components (R_L).

in the inductance or capacitance values increase the system bandwidth but change the damping ratio and natural frequency. Figure 7.25 shows that the parasitic resistance increases the damping ratio and reduces the DC gain.

In conclusion, the linearized model enables the effect of external and internal perturbation sources on the system dynamics to be taken into account. This is useful for design robust control systems and protective circuitry that will prevent undesired behavior, such as instability caused by variation in v_b. Moreover, the non-linear models, both switched and averaged, are useful for simulating the PV system and for testing large-signal controllers, such as MPPT algorithms.

In the following chapter such models are used to design control systems to ensure correct operation of the MIU.

7.5 Additional Examples

This section presents two additional modeling examples: a MIU based on a buck converter for applications where a voltage step-down is required and a MIU based on a buck–boost converter where voltage step-up/down functionality is needed. The buck-based MIU is useful for battery chargers [20] and grid-connected applications with large strings. The buck–boost based MIU is useful for applications in which the load voltage has a wider range of variation.

The examples present both switched and averaged differential equations, the state-space model, and the transfer functions of the states with respect to the control command. Such mathematical expressions can be used to extend all the analyses presented in this chapter up to now to PV systems based on buck and buck–boost converters.

7.5.1 MIU based on a Buck Converter

Using the procedure described in Section 7.2, the switched equations of the MIU in Figure 7.26 are obtained, as in (7.72) and (7.73), where $v_C = v_{pv}$. These equations also

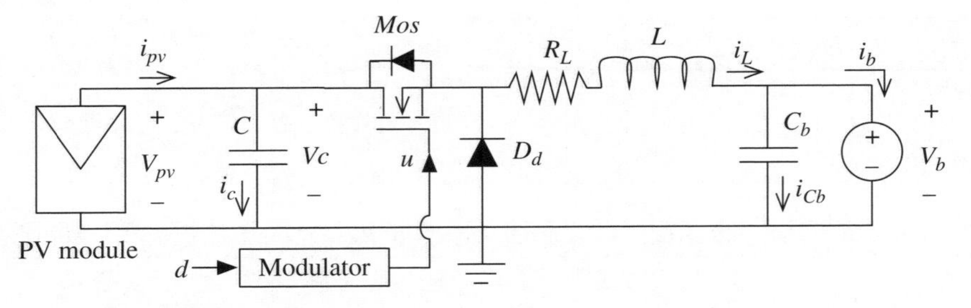

Figure 7.26 MIU based on a buck converter.

use the non-linear model of the PV module given in (7.30) to calculate the PV current i_{pv} from the module voltage.

$$L \cdot \frac{\partial i_L}{\partial t} = v_{pv} \cdot u - i_L \cdot R_L - v_b \tag{7.72}$$

$$C \cdot \frac{\partial v_{pv}}{\partial t} = i_{pv} - i_L \cdot u \tag{7.73}$$

Then, the averaged differential equations modeling the system are:

$$L \cdot \frac{\partial \overline{i_L}}{\partial t} = \overline{v_{pv}} \cdot d - \overline{i_L} \cdot R_L - v_b \tag{7.74}$$

$$C \cdot \frac{\partial \overline{v_{pv}}}{\partial t} = i_{pv} - \overline{i_L} \cdot d \tag{7.75}$$

The averaged dynamic model, represented in the state-space form by (7.17) and (7.18), is obtained following the procedure described in Section 7.2.2. The averaged state and input vectors of the MIU are defined as $\overline{\mathbf{X}} = \left[\overline{i_L}\ \overline{v_{pv}}\right]^T$ and $\mathbf{U} = \left[d\ v_b\ i_{ph}\right]^T$, respectively, and the state-space system is given in (7.76) and (7.77). This system was defined to provide information about both PV voltage and inductor current values, but to design a classical controller to regulate the PV voltage only, Equation 7.77 can be modified to $\left[\overline{v_{pv}}\right] = [0\ 1]^T \cdot \overline{\mathbf{X}}$.

$$\underbrace{\begin{bmatrix} \frac{\partial \overline{i_L}}{\partial t} \\ \frac{\partial \overline{v_{pv}}}{\partial t} \end{bmatrix}}_{\frac{\partial \overline{\mathbf{X}}}{\partial t}} = \underbrace{\begin{bmatrix} \frac{-R_L}{L} & \frac{d}{L} \\ \frac{-d}{C} & \frac{1}{C} \cdot \frac{\partial i_{pv}}{\partial v_{pv}} \end{bmatrix}}_{\mathbf{A}} \cdot \underbrace{\begin{bmatrix} \overline{i_L} \\ \overline{v_{pv}} \end{bmatrix}}_{\overline{\mathbf{X}}} + \underbrace{\begin{bmatrix} \frac{\overline{v_{pv}}}{L} & \frac{-1}{L} & 0 \\ \frac{-\overline{i_L}}{C} & 0 & \frac{1}{C} \end{bmatrix}}_{\mathbf{B}} \cdot \underbrace{\begin{bmatrix} d \\ v_b \\ i_{ph} \end{bmatrix}}_{\mathbf{U}} \tag{7.76}$$

$$\underbrace{\begin{bmatrix} \overline{i_L} \\ \overline{v_{pv}} \end{bmatrix}}_{\overline{\mathbf{Y}}} = \underbrace{\begin{bmatrix} 1 & 0 \\ 0 & 1 \end{bmatrix}}_{\mathbf{C}} \cdot \underbrace{\begin{bmatrix} \overline{i_L} \\ \overline{v_{pv}} \end{bmatrix}}_{\overline{\mathbf{X}}} + \underbrace{\begin{bmatrix} 0 & 0 & 0 \\ 0 & 0 & 0 \end{bmatrix}}_{\mathbf{D}} \cdot \underbrace{\begin{bmatrix} d \\ v_b \\ i_{ph} \end{bmatrix}}_{\mathbf{U}} \tag{7.77}$$

Finally, the dynamics of the states in the PV system of (7.76) and (7.77), with respect to the duty-cycle command, are accurately represented by the transfer functions given in (7.78) and (7.79).

$$G_{v_{pv}/d}(s)=\frac{\overline{v_{pv}}(s)}{d(s)}=\frac{-\frac{1}{C}\cdot\left(\overline{i_L}\cdot s+\frac{R_L\cdot\overline{i_L}+d\cdot v_{pv}}{L}\right)}{s^2+\left(\frac{R_L}{L}-\frac{1}{C}\cdot\frac{\partial i_{pv}}{\partial v_{pv}}\right)\cdot s+\left(\frac{d^2-R_L\cdot\frac{\partial i_{pv}}{\partial v_{pv}}}{L\cdot C}\right)} \tag{7.78}$$

$$G_{i_L/d}(s)=\frac{\overline{i_L}(s)}{d(s)}=\frac{\frac{1}{L}\cdot\left(\overline{v_{pv}}\cdot s-\frac{d\cdot\overline{i_L}+\frac{\partial i_{pv}}{\partial v_{pv}}\cdot\overline{v_{pv}}}{C}\right)}{s^2+\left(\frac{R_L}{L}-\frac{1}{C}\cdot\frac{\partial i_{pv}}{\partial v_{pv}}\right)\cdot s+\left(\frac{d^2-R_L\cdot\frac{\partial i_{pv}}{\partial v_{pv}}}{L\cdot C}\right)} \tag{7.79}$$

7.5.2 MIU based on a Buck–Boost Converter

The switched equations of the MIU in Figure 7.27 are obtained from the procedure described in Section 7.2, as in (7.80) and (7.81), where $v_C = v_{pv}$. Again, these equations use the non-linear model of the PV module given in (7.30) to calculate the PV current i_{pv} from the module voltage.

$$L\cdot\frac{\partial i_L}{\partial t}=v_{pv}\cdot u-i_L\cdot R_L+v_b\cdot(1-u) \tag{7.80}$$

$$C\cdot\frac{\partial v_{pv}}{\partial t}=i_{pv}-i_L\cdot u \tag{7.81}$$

The averaged differential equations modeling the system are given in (7.82) and (7.83), and the state-space representations (Equations 7.17 and 7.18) are given in (7.84) and (7.85). This model considers the same averaged state and input vectors as defined in the previous example.

$$L\cdot\frac{\partial\overline{i_L}}{\partial t}=\overline{v_{pv}}\cdot d-\overline{i_L}\cdot R_L+v_b\cdot(1-d) \tag{7.82}$$

$$C\cdot\frac{\partial\overline{v_{pv}}}{\partial t}=i_{pv}-\overline{i_L}\cdot d \tag{7.83}$$

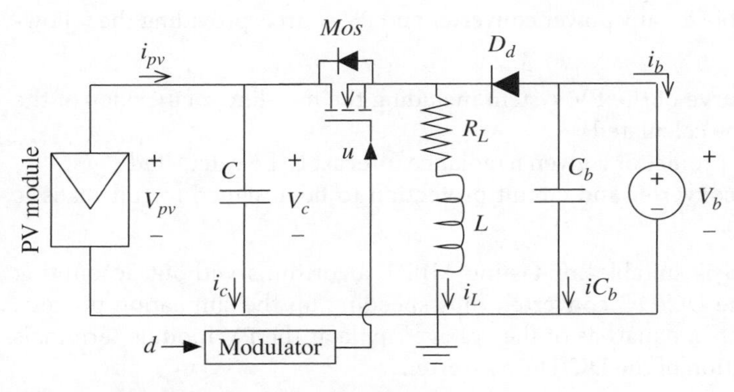

Figure 7.27 MIU based on a buck–boost converter.

$$\overbrace{\begin{bmatrix} \frac{\partial \overline{i_L}}{\partial t} \\ \frac{\partial \overline{v_{pv}}}{\partial t} \end{bmatrix}}^{\frac{\partial \overline{X}}{\partial t}} = \overbrace{\begin{bmatrix} \frac{-R_L}{L} & \frac{d}{L} \\ \frac{-d}{C} & \frac{1}{C} \cdot \frac{\partial i_{pv}}{\partial v_{pv}} \end{bmatrix}}^{\mathbf{A}} \cdot \overbrace{\begin{bmatrix} \overline{i_L} \\ \overline{v_{pv}} \end{bmatrix}}^{\overline{\mathbf{X}}} + \overbrace{\begin{bmatrix} \frac{\overline{v_{pv}} - v_b}{L} & \frac{1-d}{L} & 0 \\ \frac{-\overline{i_L}}{C} & 0 & \frac{1}{C} \end{bmatrix}}^{\mathbf{B}} \cdot \overbrace{\begin{bmatrix} d \\ v_b \\ i_{ph} \end{bmatrix}}^{\mathbf{U}} \tag{7.84}$$

$$\underbrace{\begin{bmatrix} \overline{i_L} \\ \overline{v_{pv}} \end{bmatrix}}_{\overline{\mathbf{Y}}} = \underbrace{\begin{bmatrix} 1 & 0 \\ 0 & 1 \end{bmatrix}}_{\mathbf{C}} \cdot \underbrace{\begin{bmatrix} \overline{i_L} \\ \overline{v_{pv}} \end{bmatrix}}_{\overline{\mathbf{X}}} + \underbrace{\begin{bmatrix} 0 & 0 & 0 \\ 0 & 0 & 0 \end{bmatrix}}_{\mathbf{D}} \cdot \underbrace{\begin{bmatrix} d \\ v_b \\ i_{ph} \end{bmatrix}}_{\mathbf{U}} \tag{7.85}$$

Finally, the dynamics of the states in this PV system, with respect to the duty-cycle command, are accurately represented by the transfer functions given in (7.86) and (7.87).

$$G_{v_{pv}/d}(s) = \frac{\overline{v_{pv}}(s)}{d(s)} = \frac{-\frac{1}{C} \cdot \left(\overline{i_L} \cdot s + \frac{R_L \cdot \overline{i_L} + d \cdot \overline{v_{pv}} - v_b \cdot d}{L} \right)}{s^2 + \left(\frac{R_L}{L} - \frac{1}{C} \cdot \frac{\partial i_{pv}}{\partial v_{pv}} \right) \cdot s + \left(\frac{d^2 - R_L \cdot \frac{\partial i_{pv}}{\partial v_{pv}}}{L \cdot C} \right)} \tag{7.86}$$

$$G_{i_L/d}(s) = \frac{\overline{i_L}(s)}{d(s)} = \frac{\frac{1}{L} \cdot \left(\left[\overline{v_{pv}} - v_b \right] \cdot s - \frac{d \cdot \overline{i_L} + \frac{\partial i_{pv}}{\partial v_{pv}} \cdot \left[\overline{v_{pv}} - v_b \right]}{C} \right)}{s^2 + \left(\frac{R_L}{L} - \frac{1}{C} \cdot \frac{\partial i_{pv}}{\partial v_{pv}} \right) \cdot s + \left(\frac{d^2 - R_L \cdot \frac{\partial i_{pv}}{\partial v_{pv}}}{L \cdot C} \right)} \tag{7.87}$$

7.6 Summary

This chapter has shown the impact of DC/DC power converters in the photovoltaic system. In particular, the static model of the power conversion chain has been analyzed and used for simulation purposes. Similarly, the dynamic behavior of the power conversion chain is studied in order to develop models suitable for both simulation and control.

Section 7.3 presented a simple method to model a MIU in steady-state conditions. This procedure is suitable for any power converter and PV source, providing the following features:

- It enables the I–V curve of the PV system including the non-linear efficiency of the power converter to be calculated
- It enables the power profile for a given irradiance forecast to be calculated
- It enables both the converter and circuit protection to be designed to suit realistic operating conditions.

In addition, the model is suitable for testing MPPT algorithms without accounting for the dynamics of the DC/DC converter, thus speeding up the simulation process. Section 7.3 also presents an analysis of the voltage ripple at the PV module terminals generated by the operation of the DC/DC converter.

Section 7.4 describes a procedure to model the MIU in three different ways:

- with a switched model suitable for detailed simulations and design of non-linear controllers
- with an averaged model suitable for fast simulations and design of linear and non-linear controllers
- with a linearized model for dynamic analysis and linear controller design.

The models have been used to analyze the changes of the MIU dynamics depending on the perturbations presented by the ambient conditions and load profiles. Such robustness analyses are useful for defining the optimal parameters of PV voltage controllers and MPPT algorithms.

References

1 Erickson, R.W. and Maksimovic, D. (2001) *Fundamentals of Power Electronics*, 2nd edn. Springer.

2 SolarEdge (2014) Power Optimizer. URL: http://www.solaredge.com/groups/powerbox-power-optimizer.

3 Ceraolo, M. and Poli, D. (2014) *Fundamentals of Electric Power Engineering: From Electromagnetics to Power Systems*, 1st edn. Wiley-IEEE Press.

4 Arango, E., Ramos-Paja, C.A., Calvente, J., Giral, R., and Serna-Garces, S.I. (2013) Asymmetrical interleaved dc/dc switching converters for photovoltaic and fuel cell applications-part 2: Control-oriented models. *Energies*, **6** (10), 5570–5596.

5 Vasca, F., Iannelli, L., Camlibel, M., and Frasca, R. (2009) A new perspective for modeling power electronics converters: complementarity framework. *Power Electronics, IEEE Transactions on*, **24** (2), 456–468.

6 Trejos, A., Gonzalez, D., and Ramos-Paja, C.A. (2012) Modeling of step-up grid-connected photovoltaic systems for control purposes. *Energies*, **5** (6), 1900–1926.

7 Arango, E., Ramos-Paja, C.A., Calvente, J., Giral, R., and Serna, S. (2012) Asymmetrical interleaved dc/dc switching converters for photovoltaic and fuel cell applications – Part 1: Circuit generation, analysis and design. *Energies*, **5** (11), 4590–4623.

8 Femia, N., Petrone, G., Spagnuolo, G., and Vitelli, M. (2005) Optimization of perturb and observe maximum power point tracking method. *Power Electronics, IEEE Transactions on*, **20** (4), 963–973.

9 Femia, N., Petrone, G., Spagnuolo, G., and Vitelli, M. (2009) A technique for improving P&O MPPT performances of double-stage grid-connected photovoltaic systems. *Industrial Electronics, IEEE Transactions on*, **56** (11), 4473–4482.

10 Romero-Cadaval, E., Spagnuolo, G., Garcia Franquelo, L., Ramos-Paja, C., Suntio, T., and Xiao, W. (2013) Grid-connected photovoltaic generation plants: components and operation. *Industrial Electronics Magazine, IEEE*, **7** (3), 6–20.

11 Orozco-Gutierrez, M., Ramirez-Scarpetta, J., Spagnuolo, G., and Ramos-Paja, C. (2014) A method for simulating large PV arrays that include reverse biased cells. *Applied Energy*, **123** (0), 157–167.

12 Adinolfi, G., Femia, N., Petrone, G., Spagnuolo, G., and Vitelli, M. (2010) Design of dc/dc converters for DMPPT PV applications based on the concept of energetic efficiency. *Journal of Solar Energy Engineering, ASME,* **132** (2), 021 005.

13 Petrone, G., Spagnuolo, G., Teodorescu, R., Veerachary, M., and Vitelli, M. (2008) Reliability issues in photovoltaic power processing systems. *Industrial Electronics, IEEE Transactions on,* **55** (7), 2569–2580.

14 Mamarelis, E., Ramos-Paja, C., Petrone, G., Spagnuolo, G., Vitelli, M., and Giral, R. (2013) Reducing the hardware requirements in FPGA based controllers: a photovoltaic application. *Revista Facultad de Ingenieria,* **68**, 75–87.

15 Ramos-Paja, C., Gonzalez, D., and Saavedra-Montes, A. (2013) Accurate calculation of settling time in second order systems: a photovoltaic application. *Revista Facultad de Ingenieria,* **66**, 104–117.

16 Ogata, K. (2005) *Modern Control Engineering,* 3rd edn. Prentice Hall.

17 Kuo, B. and Golnaraghi, F. (2002) *Automatic Control Systems,* 7th edn. Prentice Hall.

18 Bert, C. (1986) An improved approximation for settling time of second-order linear systems. *Automatic Control, IEEE Transactions on,* **31** (7), 642–643.

19 Piche, R. (1987) Comments on an improved approximation for settling time of second-order linear systems. *Automatic Control, IEEE Transactions on,* **32** (8), 747–748.

20 Lee, J., Bae, H., and Cho, B.H. (2008) Resistive control for a photovoltaic battery charging system using a microcontroller. *Industrial Electronics, IEEE Transactions on,* **55** (7), 2767–2775.

21 Texas Instruments (2014) SolarMagic: Solar Power Optimizer. URL: http://www.ti.com/tool/SOLARMAGIC-SOLARPOWEROPTIMIZER-REF.

8

Control of the Power Conversion Chain

8.1 Introduction

Photovoltaic systems are subject to both environmental and load perturbations. The first of these are mainly caused by changes in the irradiance and temperature. The second is caused by the interaction with the load, for example, as discussed in Chapter 7, in the case of voltage oscillations at twice the grid frequency occurring in single-phase grid-connected PV plants. Therefore, controllers are used to reject both environmental and load perturbations. In the literature, the most common techniques adopted to regulate PV systems are the classical linear controllers, based on PI or PID structures [1–4], and the non-linear controllers, based on the sliding-mode technique [5, 6].

Linear controllers are widely adopted because of their simple design and implementation. They are designed with system transfer functions and using pulse-width modulators (PWMs) to impose a fixed switching frequency. The controllers designed using this technique are able to ensure the desired performance around a given operating point, but this could put at risk the stability of power conversion chains operating far from those operating points.

Using the non-linear model of the PV power conversion chain allows more general controllers to be designed. Sliding-mode controllers (SMCs) are able to ensure the desired performance in a wider operating range. This feature makes SMCs a safer option in terms of stability, but at the price of a more complex mathematical analysis. Moreover, the classical implementation of SMCs involves a switching operation of the power converter at a variable-frequency.

The previous chapter presented mathematical models for different MIU topologies; this chapter deals with the design of controllers to regulate the PV voltage of a MIU. The general structure of a classical closed-loop MIU is presented in Figure 8.1, where a voltage controller regulates the DC/DC converter to follow a reference signal imposed by an MPPT algorithm.

The control techniques presented in this chapter are illustrated by referring to a MIU based on a boost converter – the circuit in Figure 7.6 – but the analyses can be extended to any DC/DC converter. The first approach, introduced in Section 8.2, is focused on the design of linear controllers based on the linearized model of the MIU, while the second approach, presented in Section 8.3, adopts the sliding-mode control technique, which is in turn based on the non-linear models of the MIU. Finally, Section 8.4 gives the summary of the chapter.

Photovoltaic Sources Modeling, First Edition. Giovanni Petrone, Carlos Andrés Ramos-Paja and Giovanni Spagnuolo.

Companion Website: www.wiley.com/go/petrone/Photovoltaic_Sources_Modeling

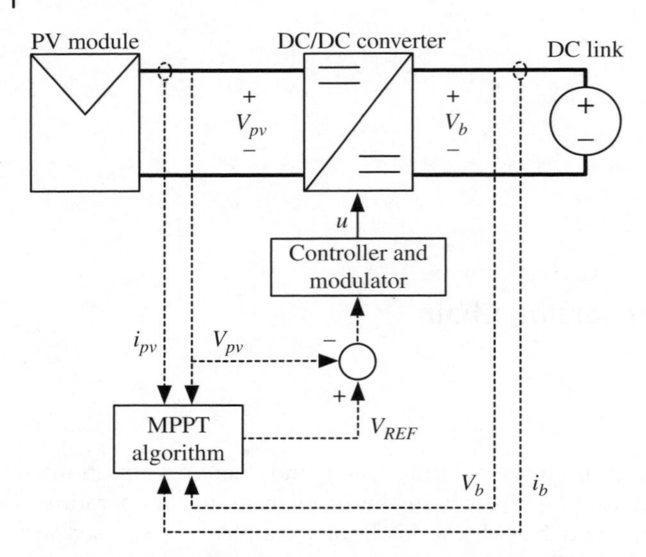

Figure 8.1 General structure of a closed-loop MIU.

8.2 Linear Controller

Classical linear controllers, such as PI and PID structures, are commonly designed using techniques based on time-response or frequency-response criteria. In general, frequency-response approaches are used to define stability criteria, which profit from systems modeling by transfer functions.

This section is focused on the design of controllers, based on both the Bode diagram and the root locus plot, that guarantee the desired gain and phase margins, closed-loop bandwidth, damping ratio, settling time, and maximum overshoot to the PV system. Such an approach leaves to the designer to impose a relative stability margin and to define the desired time response of the PV voltage.

Figure 8.2 shows the scheme of the MIU, including the control structure, where the following blocks have been added to the MIU in agreement with Figure 8.1:

- an MPPT controller, which defines the voltage reference V_{REF} followed by the PV module voltage
- a linear controller $G_v(s)$, which defines the duty cycle d of the modulator that allows the voltage reference V_{REF} to be acheived.

$G_v(s)$ must be designed to assure that the PV voltage follows V_{REF} in any environmental and load conditions: by neglecting the temperature variations, which are usually very slow, the environmental conditions are mainly represented by the photo-induced current I_{ph} in the PV module model (Equation 7.30), while the load conditions are mainly defined by the voltage v_b, which is, for instance, the DC-link voltage in a double-stage inverter or the battery voltage in a photovoltaic battery charger.

The linear controller is designed using the linear model given in (7.65) and (7.66), which takes into account I_{ph}, v_b, and d as inputs. This allows the designer to evaluate the performance of a controller in the presence of perturbations affecting both the

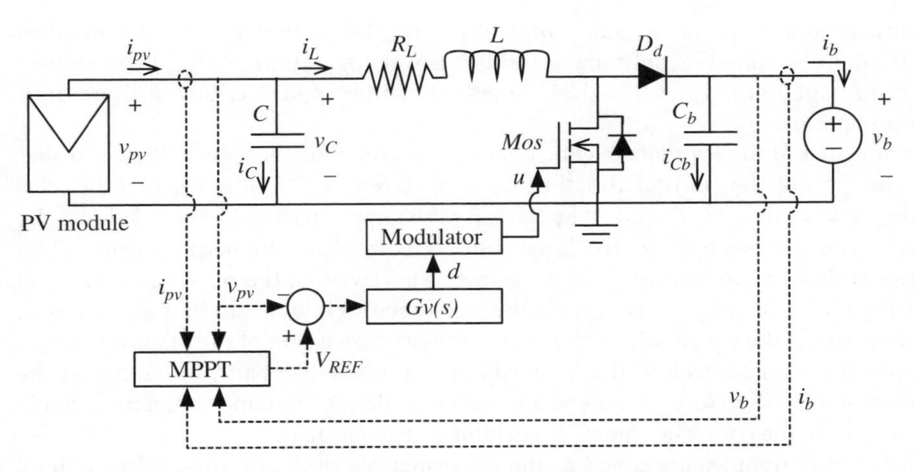

Figure 8.2 MIU based on a boost converter using linear control of the PV voltage.

environmental and load conditions. In the following, a simple method to design linear controllers for the MIU described in Figure 8.2 is presented.

The method consists of an iterative process in which the controller parameters are modified until the design criteria are fulfilled. The controller parameters are modified by changing the controller gain, poles, and zeros, while design constraints to be tested iteration after iteration are defined by the designer. The number of parameters in the controller – two parameters in PI and three parameters in PID structures – define the number of design constraints. The method consists of the following steps:

1) Add a pole in $s = 0$; that is, an integrator.
2) Add one or two zeros, for PI and PID structures respectively, and set the controller gain according to some of the following criteria:
   (a) Draw the root locus plot to verify the damping ratio of the system.
   (b) Draw the open-loop Bode diagram to verify the gain and phase margins.
   (c) Draw the closed-loop Bode diagram to verify the system bandwidth.
   (d) Draw the step response plot to verify the overshoot, settling-time, rising time, or other criteria needed.
3) If the design constraints (e.g., desired bandwidth or settling-time) are not fulfilled:
   (a) Modify the poles, zeros, and controller gain.
   (b) Add new poles or zeros if needed.
   (c) Modify the controller gain.
   (d) Go to Step 2(a).

To illustrate the design method, a MIU with the same parameters used in the previous chapter is adopted ($A = 8.95 \times 10^{-7}$ A, $B = 1.406$ V^{-1}, $C = 50\,\mu$F, $L = 100\,\mu$H, $R_L = 300$ mΩ, $F_{sw} = 50$ kHz). The MIU is considered operating under irradiance and load conditions that are $G = 500$ W/m^2 and $v_b = 24$ V. Moreover, since the MIU includes an MPPT algorithm, as in Figure 8.2, it operates at the optimal point for maximizing the power delivered to the load, which is defined by $\partial i_{pv}/\partial v_{pv} = -0.3064\,\Omega^{-1}$ and $V_{Max,b} = 8.825$ V; namely at the optimal duty cycle $D_{Max,b} = 0.661$. Such operating conditions are only taken as an example; others can be considered. For instance, the MIU in Figure 8.2

includes sensors at the input and output of the DC/DC converter, in order to allow MPPT solutions aimed at optimizing the PV power (operation at V_{MPP}) or the load power (operation at $V_{Max,b}$). The design method can be applied to any of those optimization options.

The first step of the PV voltage controller design concerns the sign of the controller gain. The gain of the transfer function $G_{v_{pv}/d}(s)$, given in (7.67), is negative because $\partial i_{pv}/\partial v_{pv} < 0$, while v_b, L, C, and R_L are positive. The upper part of Figure 8.3 shows the block diagram of a feedback control loop composed of a plant with negative gain $-G_x(s)$ in series with a controller with positive gain $G_c(s)$. Rewriting the plant as $G_x(s) \cdot (-1)$ and using block diagram equivalences, the same feedback loop can be redrawn as in the lower part of the figure, where the effect of the negative gain of the plant is evident: using positive-gain controllers the loop exhibits a positive feedback, which makes the system unstable. Therefore, to ensure a negative feedback loop and therefore a stable control system, the controller must also exhibit a negative gain.

Another important aspect concerns the performance criteria for the control system. There are multiple approaches depending on the application, focusing on relative stability, time response, and so on. One common practice in PV system controller design is to ensure safe relative stability margins and to impose a desired time response based on the MPPT requirements. For example, Gonzalez et al. impose a gain margin larger than 10 dB and a phase margin larger than 45°, which provides a safe margin to ensure stability in presence of variations of the irradiance and load voltage values [1]. Similarly, Femia et al. impose a given settling-time according to the MPPT algorithm; for example, the period of a P&O algorithm [3]. Other approaches include:

- defining the damping ratio of the complex poles to 0.707, which provides a good tradeoff between maximum overshoot and settling-time [7]
- ensuring the maximum bandwidth for the PV system that is achievable with the linearized model, say 1/10 or 1/5 of the switching frequency [2, 5, 7], which in turns provides a faster rising time.

To illustrate the controller design procedure, this section considers the following criteria:

- Ensure the largest bandwidth possible for the PV voltage: this provides a fast response to track the reference signal provided by the MPPT algorithm and to mitigate environmental and load perturbations.
- Ensure a good tradeoff between maximum overshoot and settling-time: it provides a response time balancing small overshoot and short settling time.

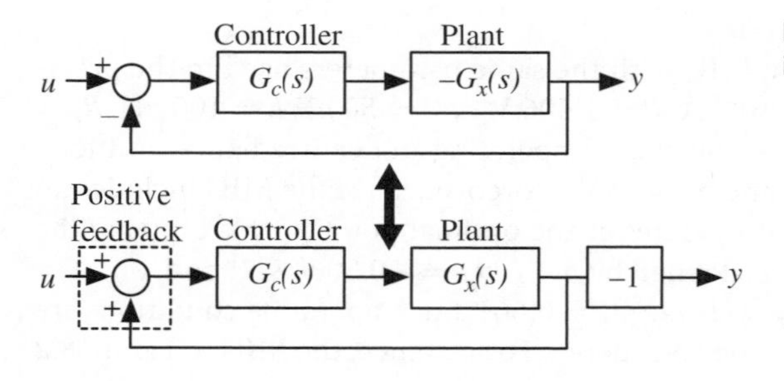

Figure 8.3 Effect of negative-gain plant in feedback control systems.

- Ensure a safe relative stability: it provides robustness to changes in the environmental and load conditions.

Since the control method requires an iterative procedure (adjust the controller parameters and verify the performance criteria), it is desirable to use an interactive tool to update, automatically, all the figures and criteria calculations when a controller parameter is changed. Such a tool can be programmed in the C language or MATLAB® scripts, but it is simpler to use the *sisotool* application available in the control system toolbox of MATLAB® [8]. *sisotool* implements the classical feedback loop (see Figure 8.3), among other control architectures. Moreover, it has a graphical interface that allows the controller parameters to be modified interactively, thus updating in real time the closed- and open-loop Bode diagrams, root-locus plot, and step response, among others.

The first step in the controller design is the calculation of the transfer function $G_{v_{pv}/d}(s)$ describing the behavior of the PV voltage in relation to the duty cycle. With this aim, the model in (7.65) and (7.66) must be linearized around the operating point of $G_{v_{pv}/d}(s)$. This transfer function is given, in the previous chapter, as Equation (7.67) for the boost-based MIU, and its evaluation around $V_{pv} = V_{Max,b} = 8.825$ V ($D = D_{Max,b} = 0.661$) is:

$$G_{v_{pv}/d}(s)\Big|_{D=0.661} = \frac{-4.8\times 10^{9}}{s^2 + 9125\cdot s + 2.184\times 10^{8}} \tag{8.1}$$

Next, *sisotool* is launched with $G_{v_{pv}/d}(s)$ as an argument, the gain of the controller is set to negative, and an integrator (pole at $s = 0$) is added. This example is focused in designing a PID controller, hence two zeros must be added. Taking into account that (8.1) exhibits two conjugate complex poles, two complex conjugate zeros are added. Figure 8.4 shows the *sisotool* interface, with both the plant $G_{v_{pv}/d}(s)$ and the PID pole and zeros. The two complex conjugate zeros and the gain of the PID are designed to achieve a damping ratio equal to 0.707 and a closed-loop bandwidth equal to 10 kHz. The upper-left figure shows the damping-ratio criterion, while the lower-left figure shows the closed-loop criterion. The right-hand figures present the open-loop Bode diagram, which is useful for verifying the relative stability of the system in terms of gain and phase margins. In addition, the step response can be plotted to tune the controller in agreement with the overshoot and settling-time requirements.

From the *sisotool*-based design, the controller $G_{pid,v_{pv}}(s)$ given in (8.2) is obtained. The performance of such a controller is first tested using the linearized model of the PV system in both the time and frequency domains. Figure 8.5 presents, on the left-hand side, the step response to changes affecting the reference voltage (T_{ref} waveform), the load voltage (T_{vb} waveform), and the irradiance (T_{io} waveform). The overshoot, settling time and null steady-state error of the PV voltage can be seen. Moreover, the time response to both load and irradiance perturbations show large attenuations, similar settling-time, and complete steady-state rejection.

$$G_{pid,v_{pv}}(s) = -\frac{1.035\times 10^{-5}\cdot s^2 + 0.2527\cdot s + 2973}{s} \tag{8.2}$$

Figure 8.5 presents, on the right-hand side, the closed-loop frequency response of the PV voltage to changes affecting the three inputs: reference voltage, load voltage, and irradiance. It shows the correct behavior of the controller in terms of the closed-loop bandwidth and null steady-state error. In addition, the satisfactory mitigation of perturbations at low frequencies is evident; the smallest mitigation of perturbations affecting

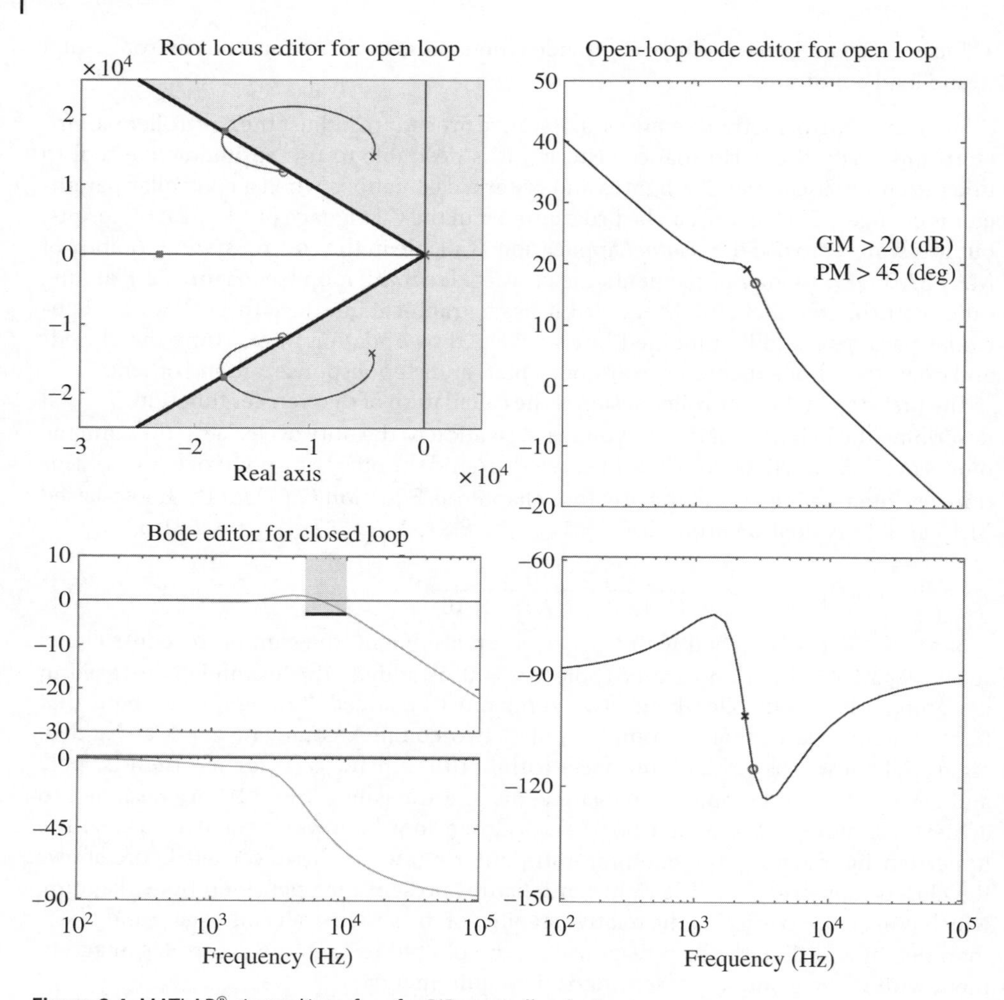

Figure 8.4 MATLAB® *sisotool* interface for PID controller design.

the load voltage is obtained at 3.2 kHz (−24.4 dB), while the smallest mitigation of perturbations in the irradiance is obtained at 4.8 kHz (−7.3 dB). Typical load voltage perturbations occur at twice the grid frequency (100 Hz in Europe and 120 Hz in the USA) where the controller ensures −50 dB attenuation. Similarly, irradiance changes are naturally slow – say 50 W/m^2 [3, 9] – and at such low frequencies the controller ensures large attenuation of those perturbations.

To test the controller, the scheme in Figure 8.2 is implemented in PSIM. The circuit simulation allows verification of the performance of the controller that operates on the real non-linear circuit: with respect to changes in the operating point and using realistic MPPT waveforms. Figure 8.6 shows the PSIM simulation with different perturbation profiles; on the left-hand side, the figure shows that the PV system satisfactorily follows a 4% change in the reference at 2 ms. Similarly, the controller satisfactorily mitigates a 10% perturbation affecting the load voltage (at 4 ms) and a 40% perturbation in the

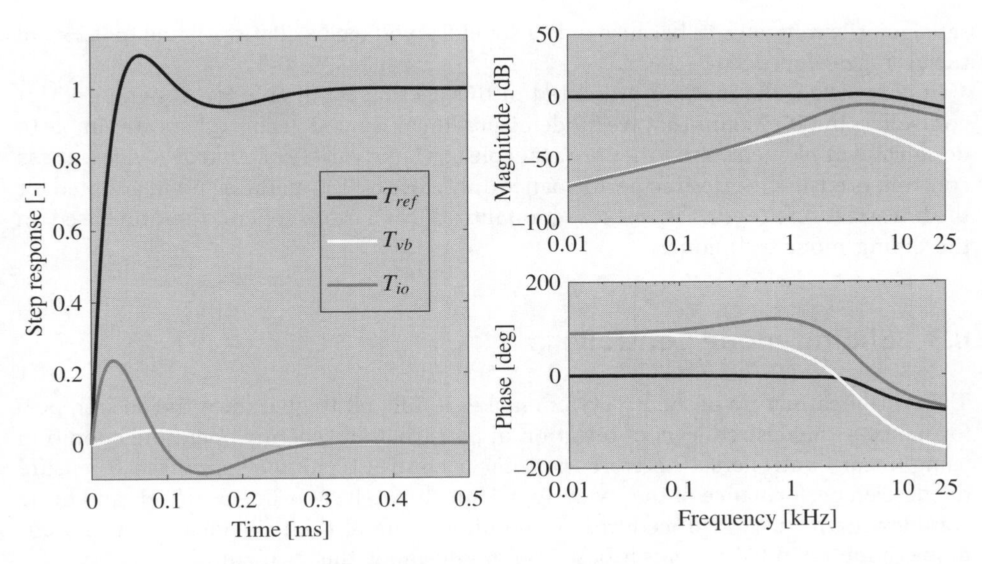

Figure 8.5 Performance of the linear controller.

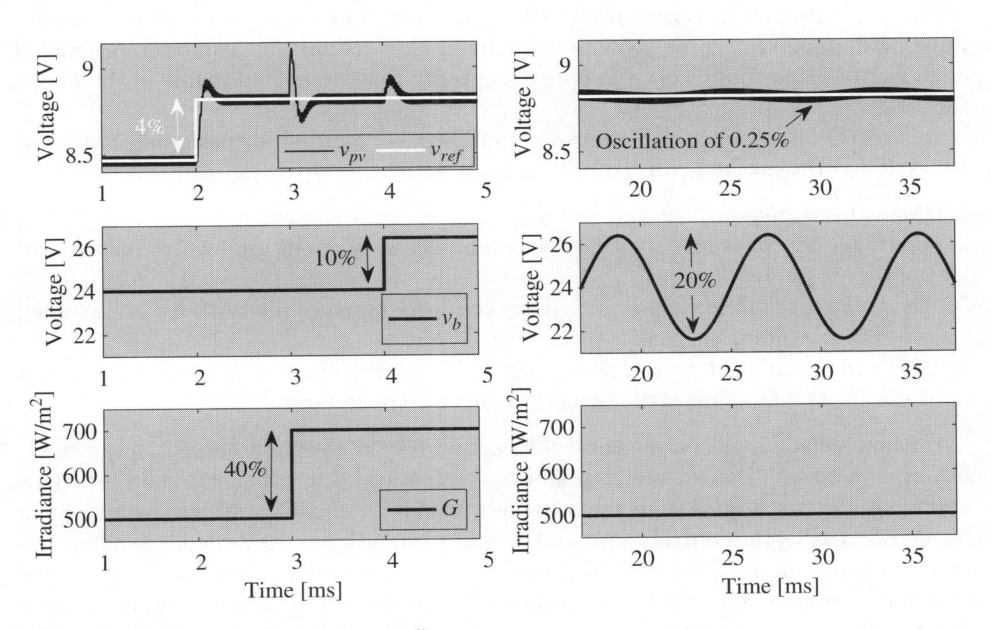

Figure 8.6 Simulation of the linear controller.

irradiance (at 3 ms). It should be noted that the waveforms predicted by the linear model in Figure 8.5 are in agreement with the circuital waveforms shown in Figure 8.6.

Figure 8.6, on the right-hand side, shows the PSIM simulation results concerning a large oscillation in the load voltage at a realistic 100 Hz frequency. This perturbation exhibits a large amplitude equal to 20% of the average voltage; the controller satisfactorily mitigates it, because only a small PV voltage oscillation, equal to 0.25% of the average value, occurs.

In conclusion, the method presented in this section is suitable for designing linear controllers in PV systems that are modeled in a linear way. Such controllers are simple to design and implement, since they are available in a large variety of control devices. These controllers ensure the desired performance only close to the operating point adopted for the linearization process. The next section introduces a more general solution based on the sliding-mode technique.

8.3 Sliding-mode Controller

The non-linear nature of the PV system makes it difficult to guarantee the desired performance – the desired level of rejection of perturbation – at any operating condition using linear controllers. Therefore, non-linear control techniques are used to ensure the desired performance of the PV system. Unfortunately, non-linear models and more complex mathematical procedures are required. One of the non-linear control techniques adopted in PV systems is based on sliding-mode theory [5, 6].

The sliding-mode technique is suitable for controlling systems based on discontinuous commands, which is the case for DC/DC converters. As a consequence, the controller must be designed using the switched equations, thus its non-linear model. A detailed analysis of sliding-mode control applied to power converters is available in the literature [10].

This section is aimed at providing a comprehensive method for designing SMCs for PV systems. The method consists of four main steps:

1) *Define the sliding surface* Ψ.
2) *Verify the transversality condition:* this is a necessary condition for the existence of the sliding mode.
3) *Verify the reachability conditions:* these conditions ensure the system will be driven towards the sliding surface.
4) *Verify the equivalent control condition:* this condition ensures local stability, so that when the system reaches the surface it will be trapped there.

The first step depends on the control objective: regulate voltage, current, impedance, power, and so on. The second step allows verification of whether the sliding surface defined in the first step is achievable; in particular, it checks if the system behavior can be modified using the control command. The third step allows the designer to detect the conditions that must be fulfilled to reach the surface, which is useful for design of the implementation circuit. The last step allows the designer to detect the conditions that must be fulfilled to guarantee the operation within the sliding surface, which is useful for detection of the constraints of the control technique, such as the maximum derivative of the reference signal, the maximum derivative of the perturbations that could be rejected, and so on. The mathematical formulations of the four steps are illustrated in the following by considering different control objectives for the MIU based on the boost converter.

Based on the same approach considered in the previous section – consisting of an MPPT algorithm that provides the voltage reference – the SMC replaces both the linear controller $Gv(s)$ and the PWM in the scheme of Figure 8.2. Then, the control objective is to regulate the PV voltage v_{pv} to track the reference voltage v_{REF}, which leads to the sliding surface given in (8.3). This is the first step of the method:

$$\Psi_{v_{pv}} = v_{pv} - v_{REF} = 0 \tag{8.3}$$

From the switched differential equations of the MIU given in the previous chapter (7.61) and (7.62), the derivative of the surface $\Psi_{v_{pv}}$ for a constant reference is the following:

$$\frac{\partial \Psi_{v_{pv}}}{\partial t} = \frac{\partial v_{pv}}{\partial t} = \frac{i_{pv} - i_L}{C} \tag{8.4}$$

A surface Ψ fulfills the transversality condition if (8.5) is true [10, 11]. This is the second step of the method. In practice, (8.5) verifies that the surface derivative depends on the control variable u.

$$\frac{\partial \frac{\partial \Psi}{\partial t}}{\partial u} \neq 0 \tag{8.5}$$

From Equation 8.4, Equation 8.6 – which shows that $\Psi_{v_{pv}}$ does not fulfill the condition given in (8.5) – is obtained. Therefore, the surface $\Psi_{v_{pv}}$ is not suitable for designing a SMC for the MIU.

$$\frac{\partial \frac{\partial \Psi_{v_{pv}}}{\partial t}}{\partial u} = 0 \tag{8.6}$$

The previous condition implies that other sliding-surfaces must be proposed. In the literature, several surfaces have been used to successfully control a PV system: one aimed at regulating the inductor current [5] and one aimed at regulating the capacitor current [6]. Both approaches are presented below.

8.3.1 Inductor Current Control

Because the dynamics of the relationship between the voltage and current of a PV module are much faster than those of the switching converter, regulation of the PV current also enables the PV voltage to be regulated. Moreover, because the steady-state current of the input capacitor is zero for the MIU, according to the charge balance principle [4], the steady-state values of both PV current and inductor current are the same. Therefore, when an SMC fixes a desired inductor current, the perturbations in both the load voltage and irradiance are mitigated.

Figure 8.7 shows the scheme used to control the PV voltage by means of an SMC regulating the inductor current. This solution has an additional controller $G_{CVL}(s)$, which defines the reference $i_{L,ref}$ for the inductor current aimed to track the voltage reference v_{REF} provided by the MPPT algorithm.

Following the method outlined in the first part of Section 8.3, the first step is to define the sliding surface. Since the control objective is to regulate the inductor current i_L of the MIU, the surface Ψ_{i_L} is defined as:

$$\Psi_{i_L} = i_L - i_{L,ref} = 0 \tag{8.7}$$

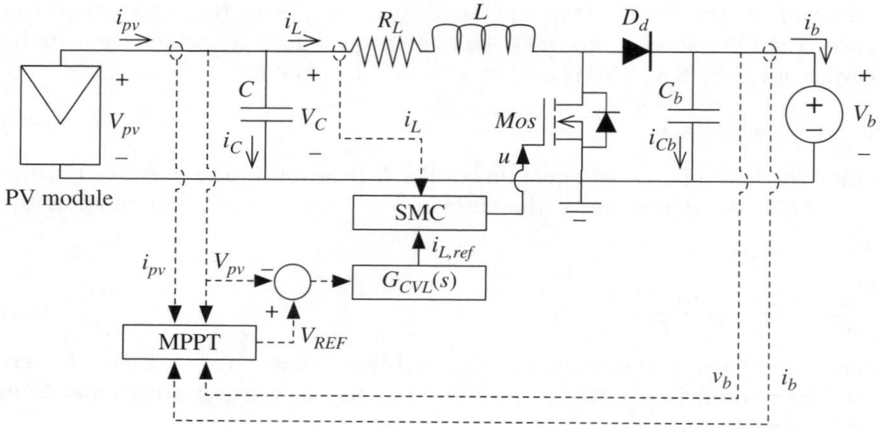

Figure 8.7 Scheme of a PV voltage SMC based on inductor current control.

From the switched differential equations of the MIU given in (7.61) and (7.62), the derivative of the surface Ψ_{i_L} for a constant reference is:

$$\frac{\partial \Psi_{i_L}}{\partial t} = \frac{\partial i_L}{\partial t} = \frac{v_{pv} - i_L \cdot R_L - v_b \cdot (1-u)}{L} \tag{8.8}$$

The second step of the method is to verify the transversality condition. From Equation 8.8, condition (8.9) is obtained, which shows that Ψ_{i_L} fulfills the condition given in (8.5). Therefore, the surface Ψ_{i_L} fulfills the transversality condition, and hence it is suitable for designing an SMC for the MIU.

$$\frac{\partial \frac{\partial \Psi_{i_L}}{\partial t}}{\partial u} = \frac{v_b}{L} > 0 \quad \Rightarrow \quad \frac{\partial \frac{\partial \Psi_{i_L}}{\partial t}}{\partial u} \neq 0 \tag{8.9}$$

It is worth noting that the transversality condition (Equation 8.9) is positive, hence the reachability conditions given in (8.10) must be granted [11, 12].

$$\left.\begin{array}{ll} \lim\limits_{\Psi_{i_L} \to 0^-} \dfrac{\partial \Psi_{i_L}}{\partial t} > 0 \rightarrow & u = 1 \\ \lim\limits_{\Psi_{i_L} \to 0^+} \dfrac{\partial \Psi_{i_L}}{\partial t} < 0 \rightarrow & u = 0 \end{array}\right\} \tag{8.10}$$

When the surface (8.7) is negative, it means the inductor current is lower than the reference current, so the inductor current must be increased. This requires a positive derivative of the inductor current, which means also a positive surface derivative (8.8). In order to obtain a positive value of the surface derivative it is required that $u = 1$. Similarly, when the surface is positive, it means the inductor current is higher than the reference current, so the inductor current must be decreased. This requires a negative derivative of the inductor current, which means also a negative surface derivative; to obtain a negative value of the surface derivative it is required that $u = 0$.

The previous analysis is used to perform the third step of the method, which consists of verifing the reachability conditions. By substituting (8.8) into (8.10), the expressions in (8.11) are obtained. These results verify the reachability conditions, so the sliding surface can be reached from any operating point: if the inductor current is lower than the reference current, the SMC will increase the inductor current by setting the MOSFET ON; if the inductor current is higher than the reference current, the SMC will decrease the inductor current by setting the MOSFET OFF.

$$\left.\begin{aligned} \lim_{\Psi_{i_L}\to 0^-} \frac{\partial \Psi_{i_L}}{\partial t} &= \frac{v_{pv} - i_L \cdot R_L}{L} > 0 \quad \text{for} \quad u = 1 \\ \lim_{\Psi_{i_L}\to 0^+} \frac{\partial \Psi_{i_L}}{\partial t} &= \frac{\left(v_{pv} - v_b\right) - i_L \cdot R_L}{L} < 0 \text{ for} \quad u = 0 \end{aligned}\right\} \tag{8.11}$$

The final step of the method is to verify the equivalent control condition. To fulfill this principle the average value u_{eq} of the control variable u must fall within the correct operating range of the system [11, 12]. For a DC/DC converter, the minimum and maximum values of u are 0 and 1, respectively, so the correct range for the equivalent control is $0 < u_{eq} < 1$.

Moreover, since the equivalent control condition concerns the local stability of the SMC, the system is analyzed within the surface, that is, $\Psi_{i_L} = 0$. Then, (8.7) leads to $i_L = i_{L,ref}$ and by replacing u by u_{eq} in (7.61), the equivalent control value is given by (8.12). Then, from the equivalent control principle, the maximum derivative of the reference that ensures the existence of the sliding-mode is obtained in (8.13).

$$0 < u_{eq} = \frac{L}{v_b} \cdot \frac{\partial i_{L,ref}}{\partial t} + 1 - \frac{v_{pv} - i_{L,ref} \cdot R_L}{v_b} < 1 \tag{8.12}$$

$$\frac{v_{pv} - i_{L,ref} \cdot R_L - v_b}{L} < \frac{\partial i_{L,ref}}{\partial t} < \frac{v_{pv} - i_{L,ref} \cdot R_L}{L} \tag{8.13}$$

It must be pointed out that the limits in (8.13) correspond to the slopes of the inductor current, for both $u = 0$ and $u = 1$, respectively, at $i_L = i_{L,ref}$. This condition shows one of the main advantages of the SMC over a linear controller: the change in the inductor current, so that the desired PV voltage is achieved, occurs in the shortest possible time. However, the slope of the reference must be constrained as in (8.13) to guarantee that the system does not operate outside of the sliding surface. For example, $i_{L,ref}$ must not exhibit step-like changes. Finally, when the system is operating in sliding mode the following conditions are fulfilled [10–12]:

$$\Psi_{i_L} = 0, \qquad \frac{\partial \Psi_{i_L}}{\partial t} = 0 \tag{8.14}$$

Another important aspect concerns the implementation of the SMC, which defines the switching frequency of the semiconductors [5, 10]. Classical implementations of SMCs gives rise to variable switching frequencies [10–12], but it is important to define the average frequency in the desired operating range to avoid incorrect operation of the system. In this way, (7.49) describes the ripple magnitude Δi_L of the inductor current in terms of the PV voltage, average inductor current, duty cycle, switching period T_{sw}, and inductance. Therefore, if the inductor current is always equal to the reference, it means that $\Delta i_L = 0$, which in turns requires $T_{sw} = 0$, and hence an infinite switching frequency F_{sw}.

The switching frequency of the SMC is constrained by imposing an hysteresis around the sliding surface in which the system can operate. In this way, the sliding surface Ψ_{i_L} in (8.7) is implemented as in (8.15), where the magnitude of the hysteresis band $\Delta\Psi_{i_L}$ affects the switching frequency. $\Delta\Psi_{i_L}$ denotes the maximum deviation of the inductor current with respect to the reference current (both positive and negative), which in practice stands for the ripple of the inductor current Δi_L. Then, replacing $\Delta\Psi_{i_L}$ by Δi_L and expanding (8.15) to remove the absolute value, the expression in (8.16) is obtained. This inequality, and the previous analysis of (8.11), provide the information required to design the implementation circuit: if Ψ_{i_L} is lower than the lower limit of the hysteresis band $\left(-\Delta i_L/2\right)$, the inductor current must be increased. This, according to the analysis of (8.11), requires the MOSFET to be turned ON. If Ψ_{i_L} is higher than the upper limit of the hysteresis band $\left(\Delta i_L/2\right)$, the inductor current must be reduced, which according to the analysis of (8.11) requires the MOSFET to be turned OFF. This behavior is summarized in (8.17).

$$\left|\Psi_{i_L} = i_L - i_{L,ref}\right| < \frac{\Delta\Psi_{i_L}}{2} \tag{8.15}$$

$$-\frac{\Delta i_L}{2} < i_L - i_{L,ref} < \frac{\Delta i_L}{2} \tag{8.16}$$

$$\left.\begin{array}{lll} i_L - i_{L,ref} < -\frac{\Delta i_L}{2} & \Rightarrow & \text{Turn ON the MOSFET } (u = 1) \\ i_L - i_{L,ref} > \frac{\Delta i_L}{2} & \Rightarrow & \text{Turn OFF the MOSFET } (u = 0) \end{array}\right\} \tag{8.17}$$

The implementation of (8.17) is performed using a subtractor to calculate Ψ_{i_L}, two comparators, and a Flip-Flop S-R, but other structures can be adopted. In such a circuit, depicted in Figure 8.8, the subtractor and comparators implement the switching laws defined in (8.17), while the Flip-Flop receives the set (turn ON) and reset (turn OFF) signals to hold the MOSFET status by means of the signal u.

Then, the inductor current magnitude Δi_L in (7.49) is granted by the SMC. This condition enables both the duty cycle d and switching frequency F_{sw} of the converter to be estimated from the model given in (7.34) and (7.49), as reported in (8.18). Then, the maximum switching frequency is achieved at the higher load voltage v_b and lower PV current. Based on this information, the value of Δi_L must be defined to achieve the desired range of the switching frequency.

$$F_{sw} \approx \frac{\left(v_{pv} - i_{pv} \cdot R_L\right)}{L \cdot \Delta i_L} \cdot d, \quad d \approx 1 - \frac{v_{pv} - i_{pv} \cdot R_L}{v_b} \tag{8.18}$$

Adopting the parameters used in the examples of the previous section, but considering a larger irradiance range between 600 and 800 W/m^2, a switching frequency close to

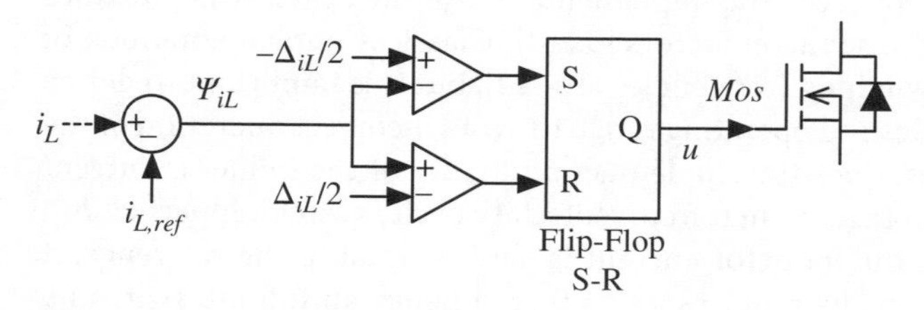

Figure 8.8 Implementation of the SMC for Ψ_{i_L}.

60 kHz for a PV current equal to 2 A can be defined by setting $\Delta i_L = 0.9635\ A \approx 1$ A. Figure 8.9 shows the simulation results of the SMC implementing Ψ_{i_L} with the scheme described in Figure 8.8, where the reference current changes from 2 A to 1 A (50%) at 6 ms, 20% and 40% perturbations in v_b are introduced at 3 ms and 5 ms, respectively, and 33% perturbations affect the irradiance at 2 ms and 4 ms. These waveforms show the excellent tracking of the reference imposed on the SMC and, at the same time, the excellent rejection of both load and ambient perturbations. The simulation waveforms also show the variation of v_{pv} due to the changes in irradiance, so an additional controller must be introduced to regulate the PV voltage in agreement with an MPPT command.

To design a voltage regulator, the dynamics of the whole PV system, including the SMC, must be modeled. From the conditions in (8.14), the PV system dynamics in (7.61) and (7.62) are changed as in (8.19) due to the SMC action. Moreover, the PV current can be expressed in terms of the PV voltage and the instantaneous PV admittance, that is $i_{pv} = \frac{\partial i_{pv}}{\partial v_{pv}} \cdot v_{pv}$. This instantaneous PV admittance is studied in Section 7.4 and analyzed in Figure 7.19 to solve the averaged model (7.65). Then, the closed-loop dynamics of the PV voltage can be described in the Laplace domain since, as observed in (8.19), it exhibits an equivalent linear behavior. The transfer function between PV voltage and inductor current reference is given in (8.20), which is always stable since $\frac{\partial i_{pv}}{\partial v_{pv}} < 0$, as depicted in Figure 7.19.

$$i_L = i_{L,ref}, \qquad \frac{\partial v_{pv}}{\partial t} = \frac{i_{pv} - i_{L,ref}}{C} \tag{8.19}$$

$$G_{v_{pv}/i_{L,ref}}(s) = \frac{V_{pv}(s)}{I_{L,ref}(s)} = -\frac{1}{s \cdot C - \frac{\partial i_{pv}}{\partial v_{pv}}} \tag{8.20}$$

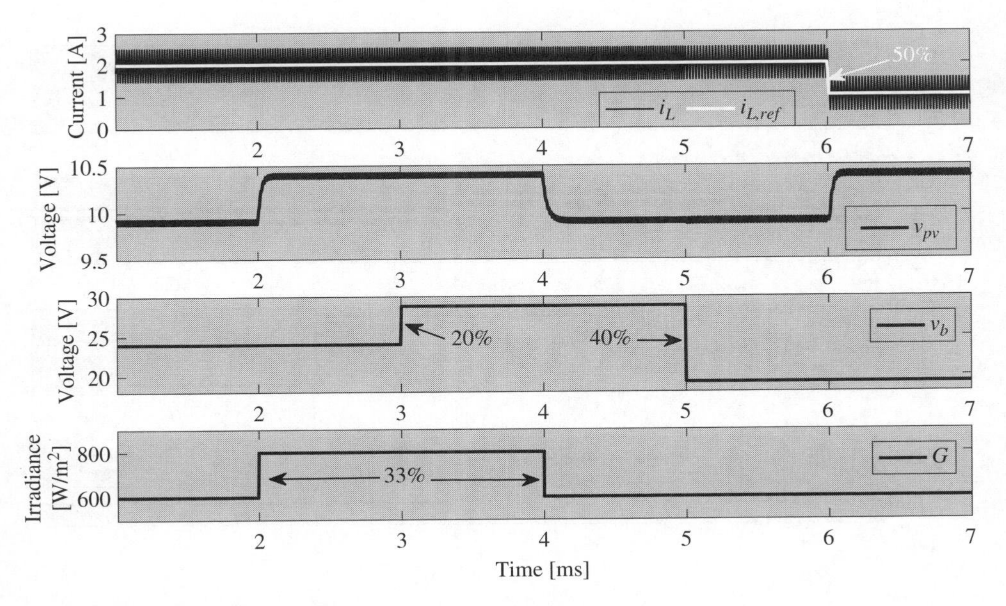

Figure 8.9 Simulation of the SMC for Ψ_{i_L}.

That first-order transfer function must be parameterized at a desired operating point by means of $\frac{\partial i_{pv}}{\partial v_{pv}}$, the value of which can be extracted from Figure 7.19. Then, the numerical version of (8.20) can be used to design a linear (or non-linear) controller $G_{CVL}(s)$ to regulate the PV voltage acting on the reference of the SMC, as depicted in Figures 8.7 and 8.8. Similar to the single linear controller presented in Section 8.2, the reference for the PV voltage is provided by an MPPT controller that is aimed at maximizing the generated power.

To illustrate the design of $G_{CVL}(s)$, the numerical values adopted for the previous example are used, where $\frac{\partial i_{pv}}{\partial v_{pv}} = -0.3867\ \Omega^{-1}$ was obtained for an average irradiance of 750 W/m^2 and an MPP voltage equal to 8.987 V. The numerical version of (8.20) evaluated at this operating point is given in (8.21). Then, $G_{CVL}(s)$ in (8.22) is designed for the criteria adopted in the previous section: damping ratio equal to 0.707 and closed-loop bandwidth equal to 10 kHz.

$$G_{v_{pv}/i_{L,ref}}(s) = -\frac{1}{5\times 10^{-5}\cdot s + 0.3867} \tag{8.21}$$

$$G_{CVL}(s) = -\frac{2.19\cdot s + 66370}{s} \tag{8.22}$$

Figure 8.10 shows the simulation results of the complete SMC-PI controller regulating the PV voltage, as in the schemes depicted in Figures 8.7 and 8.8. This simulation considers the same large perturbations exhibited in the previous example, but in this case the PV voltage is regulated by the linear controller. It must be pointed out that load perturbations, such as the 100 or 120-Hz oscillations generated in single-phase grid-connected PV systems, are completely rejected. Perturbations in the irradiance are well mitigated, while the reference voltage (such as the one provided by an MPPT algorithm) is accurately tracked.

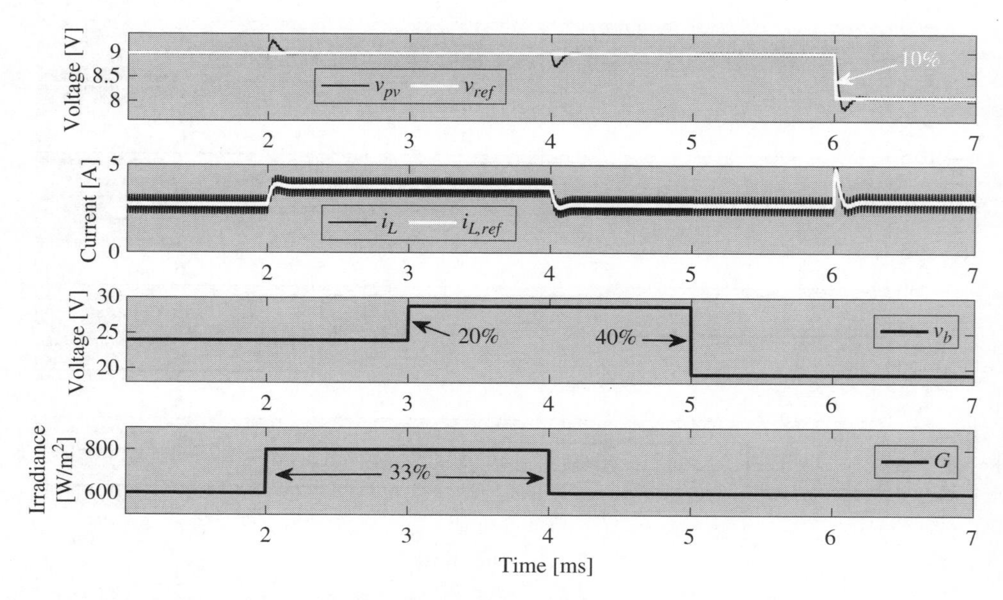

Figure 8.10 Simulation of PV voltage control including the SMC for Ψ_{i_L}.

Finally, this PV voltage regulator based on the SMC Ψ_{i_L} gives the system stability and provides improved performance over the single/PWM-based PID regulator, both in terms of perturbation rejection and reference tracking. Unfortunately, since the voltage regulator requires the value of $\frac{\partial i_{pv}}{\partial v_{pv}}$, which has to be calculated for a particular operating point, the control performance is not the same for the whole operating range. In order to overcome this limitation, the next subsection introduces an alternative SMC, based on a different sliding surface.

8.3.2 Capacitor Current Control

Another approach to regulating the PV voltage is to control the input capacitor current, which corresponds to the PV voltage derivative. Due to the charge balance principle, the average value of the input capacitor current must be zero. Then, the surface Ψ_{i_C} given in (8.23) is defined to implement an SMC to mitigate perturbations in both the load voltage and irradiance.

$$\Psi_{i_C} = i_C - i_{C,ref} = 0 \tag{8.23}$$

Figure 8.11 shows the scheme used to control the PV voltage by means of the new SMC regulating the capacitor current. Similarly to the previous SMC controller, this solution considers an additional controller $G_{CVC}(s)$ to define the reference $i_{C,ref}$ in order to track the voltage reference v_{REF} provided by the MPPT algorithm.

On the basis of the switched differential equations of the MIU given in (7.61) and (7.62), and by taking into account that $i_C = i_{pv} - i_L$, the derivative of the surface Ψ_{i_C} for a constant reference is:

$$\frac{\partial \Psi_{i_C}}{\partial t} = \frac{\partial i_C}{\partial t} = \frac{\partial i_{pv}}{\partial t} - \frac{\partial i_L}{\partial t} = \frac{\partial i_{pv}}{\partial v_{pv}} \cdot \frac{\partial v_{pv}}{\partial t} - \frac{\partial i_L}{\partial t} \tag{8.24}$$

$$\frac{\partial \Psi_{i_C}}{\partial t} = \frac{\partial i_{pv}}{\partial v_{pv}} \cdot \frac{i_C}{C} - \frac{v_{pv} - i_L \cdot R_L - v_b \cdot (1-u)}{L} \tag{8.25}$$

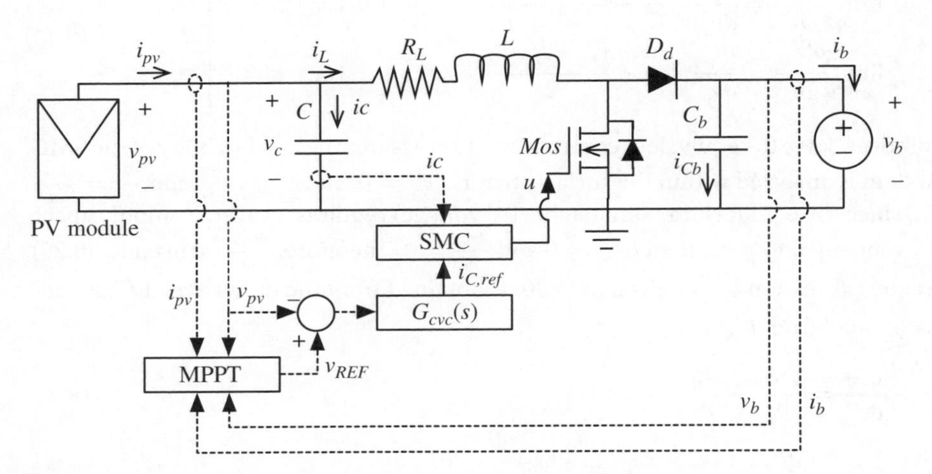

Figure 8.11 Scheme of a PV voltage SMC based on capacitor current control.

Then, (8.26) is obtained from (8.25), which demonstrates that Ψ_{i_C} fulfills the transversality condition. In addition, since the transversality condition of (8.26) is negative, the reachability conditions given in (8.27) must be fulfilled [11, 12]. These expressions mean that:

- when the surface (8.23) is negative, the capacitor current must be increased, it requiring a positive derivative of the capacitor current, which in turns requires that $u = 0$
- when the surface is positive, the capacitor current must be decreased, it requiring a negative derivative of the capacitor current, which in turn requires that $u = 1$.

$$\frac{\partial \frac{\partial \Psi_{i_C}}{\partial t}}{\partial u} = -\frac{v_b}{L} < 0 \quad \Rightarrow \quad \frac{\partial \frac{\partial \Psi_{i_C}}{\partial t}}{\partial u} \neq 0 \tag{8.26}$$

$$\left.\begin{array}{ll} \lim\limits_{\Psi_{i_C} \to 0^-} \dfrac{\partial \Psi_{i_C}}{\partial t} > 0 \to & u = 0 \\ \lim\limits_{\Psi_{i_C} \to 0^+} \dfrac{\partial \Psi_{i_C}}{\partial t} < 0 \to & u = 1 \end{array}\right\} \tag{8.27}$$

Taking into account that a stable PV voltage requires an average input capacitor current equal to zero, the reference capacitor current exhibits $i_{C,ref} = 0$ for the largest part of the operation time, hence $\Psi_{i_C} < 0 \to i_C < 0$ and $\Psi_{i_C} > 0 \to i_C > 0$ also for the largest part of the operation time. Therefore, since $\frac{\partial i_{pv}}{\partial v_{pv}} < 0$, substituting (8.25) into (8.27) leads to the expressions in (8.28), which verify the reachability conditions. These results ensure that the sliding surface is reached from any operating point: if the capacitor current is lower than the reference current, the SMC increases the capacitor current by setting the MOSFET OFF; if the capacitor current is higher than the reference current, the SMC decreases the capacitor current by setting the MOSFET ON.

$$\left.\begin{array}{ll} \lim\limits_{\Psi_{i_C} \to 0^-} \dfrac{\partial \Psi_{i_C}}{\partial t} = \dfrac{\partial i_{pv}}{\partial v_{pv}} \cdot \dfrac{i_C}{C} - \dfrac{(v_{pv} - v_b) - i_L \cdot R_L}{L} > 0 \text{ for} & u = 0 \\ \lim\limits_{\Psi_{i_C} \to 0^+} \dfrac{\partial \Psi_{i_C}}{\partial t} = \dfrac{\partial i_{pv}}{\partial v_{pv}} \cdot \dfrac{i_C}{C} - \dfrac{v_{pv} - i_L \cdot R_L}{L} < 0 \quad \text{for} & u = 1 \end{array}\right\} \tag{8.28}$$

Moreover, since the equivalent control condition assures the local stability of the SMC, the system is analyzed within the surface; that is, $\Psi_{i_C} = 0$, and $i_C = i_{C,ref}$ and $\frac{\partial i_C}{\partial t} = \frac{\partial i_{C,ref}}{\partial t}$. Again, since the equilibrium point (stable PV voltage) requires an average input capacitor current equal to zero, then $i_{C,ref} = 0$ and $\frac{\partial v_{PV}}{\partial t} = 0$. Therefore, $\frac{\partial i_{C,ref}}{\partial t}$ is given by (8.29), where the expression of $\frac{\partial i_{PV}}{\partial t}$ given in (8.30) is obtained from the derivative of (7.30) and $\frac{\partial i_L}{\partial t}$ is obtained from (7.61).

$$\frac{\partial i_{C,ref}}{\partial t} = \frac{\partial i_{PV}}{\partial t} - \frac{\partial i_L}{\partial t} \tag{8.29}$$

$$\frac{\partial i_{PV}}{\partial t} = \frac{\partial I_{ph}}{\partial t} - A \cdot B \cdot e^{(B \cdot v_{pv})} \cdot \overbrace{\frac{\partial v_{PV}}{\partial t}}^{0} = \frac{\partial I_{ph}}{\partial t} \tag{8.30}$$

Replacing u by u_{eq} in $\frac{\partial i_{C,ref}}{\partial t}$, the equivalent control value given by (8.31) is calculated. Then, from the equivalent control principle, the interaction of the reference and irradiance derivatives must be bounded, as in (8.32), to guarantee the existence of the sliding-mode ($I_{ph} = K_G \cdot G$, where K_G is a constant of proportionality between the irradiance and the photoinduced current, which can be considered as a constant value [13]).

$$0 < u_{eq} = \frac{L}{v_b} \cdot \left(\frac{\partial I_{ph}}{\partial t} - \frac{\partial i_{C,ref}}{\partial t} \right) + 1 - \frac{v_{pv} - i_L \cdot R_L}{v_b} < 1 \tag{8.31}$$

$$\frac{v_{pv} - i_L \cdot R_L - v_b}{L} < \frac{\partial I_{ph}}{\partial t} - \frac{\partial i_{C,ref}}{\partial t} < \frac{v_{pv} - i_L \cdot R_L}{L} \tag{8.32}$$

It must be pointed out that the limits in (8.32) are equal to the slopes of the inductor current for both $u = 0$ and $u = 1$, respectively, in any condition. Hence, as in the previous SMC, the SMC imposes the fastest change possible to the capacitor current, so that the desired PV voltage is achieved faster than when using the linear controller. Using the separability criterion, which assumes that fast perturbations in both I_{ph} and $i_{C,ref}$ are not concurrent [6], the maximum slope of perturbations in I_{ph} that the SMC can reject are constrained by the limits of (8.32). Similarly, to guarantee the system operation within the sliding surface, the slope of $i_{C,ref}$ must be bounded by such dynamic limits. Finally, when the system is operating in sliding mode the following conditions are fulfilled [10–12]:

$$\Psi_{i_C} = 0, \quad \frac{\partial \Psi_{i_C}}{\partial t} = 0 \tag{8.33}$$

To develop a practical SMC, the switching frequency of the SMC is bounded by implementing the sliding surface Ψ_{i_C} in (8.23) as in (8.34). In this case $\Delta\Psi_{i_C}$ denotes the maximum deviation of the capacitor current with respect to the reference current, which in practice stands for the ripple of the capacitor current Δi_C. Then, replacing $\Delta\Psi_{i_C}$ by Δi_C into (8.34) leads to (8.35). Both (8.28) and (8.35) provide the information required to design the implementation circuit of the SMC as follows: if Ψ_{i_C} is lower than the lower limit of the hysteresis band $(-\Delta i_C/2)$, the analysis of (8.28) means that the MOSFET must be turned OFF. Instead, if Ψ_{i_C} is higher than the upper bound of the hysteresis band $(\Delta i_C/2)$, the analysis of (8.28) means that the MOSFET must be turned ON. This behavior is summarized in (8.36) and its implementation circuit is presented in Figure 8.12.

$$\left| \Psi_{i_C} = i_C - i_{C,ref} \right| < \frac{\Delta\Psi_{i_C}}{2} \tag{8.34}$$

$$-\frac{\Delta i_C}{2} < i_C - i_{C,ref} < \frac{\Delta i_C}{2} \tag{8.35}$$

$$\left. \begin{array}{lll} i_C - i_{C,ref} < -\frac{\Delta i_C}{2} & \Rightarrow & \text{Turn OFF the MOSFET } (u = 0) \\ i_C - i_{C,ref} > \frac{\Delta i_C}{2} & \Rightarrow & \text{Turn ON the MOSFET } (u = 1) \end{array} \right\} \tag{8.36}$$

As for estimating the switching frequency, it should be noted that almost all the inductor current ripple flows through the input capacitor as analyzed in Section 7.3.3 (Equation 7.44), which in practice leads to $\Delta i_C \approx \Delta i_L$. In addition, since the capacitor current magnitude is imposed by the SMC, the inductor current ripple is fixed. Therefore, both the duty cycle d and switching frequency F_{sw} of the converter are estimated

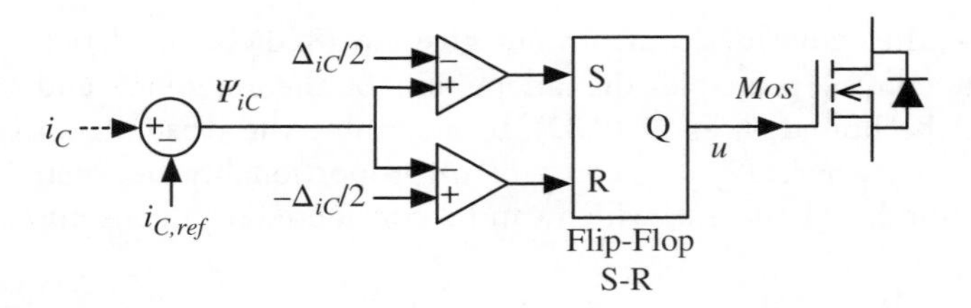

Figure 8.12 Implementation of the SMC for Ψ_{i_C}.

using the same expression (8.18) obtained for the previous SMC based on inductor current regulation.

The next step is to design a regulator for the PV voltage, where the dynamics of the PV system including the SMC must be modeled. From the conditions in (8.33), the PV voltage dynamics in (7.62) are changed as in (8.37), which is a linear relation that leads to the Laplace representation in (8.38). This model highlights an advantage of Ψ_{i_C} over Ψ_{i_L} and linear controllers: the transfer function of the PV voltage does not need to be parameterized at any particular operating point, so it guarantees the same performance of the PV voltage under any load or environmental condition.

$$\frac{\partial v_{pv}}{\partial t} = \frac{i_{C,ref}}{C} \tag{8.37}$$

$$G_{v_{pv}/i_{C,ref}}(s) = \frac{V_{pv}(s)}{I_{C,ref}(s)} = \frac{1}{s \cdot C} \tag{8.38}$$

A numerical version of (8.38) can be used to design a linear (or non-linear) controller $G_{CVC}(s)$ to regulate the PV voltage acting on the reference of the SMC, as depicted in Figures 8.11 and 8.12. As for the controllers presented in the previous sections, the reference for the PV voltage is provided by an MPPT controller to maximize the generated power.

To illustrate the design of $G_{CVC}(s)$, the numerical values adopted for the previous example – $\Delta i_L = 1$ A and MPP voltage equal to 8.987 V – are selected to give a fair comparison with the SMC based on Ψ_{i_L}. The numerical version of (8.38) is given in (8.39), and $G_{CVC}(s)$ in (8.40) is designed by accounting for the same constraints adopted in the previous subsection: damping ratio equal to 0.707 and closed-loop bandwidth equal to 10 kHz.

$$G_{v_{pv}/i_{C,ref}}(s) = \frac{1}{5 \times 10^{-5} \cdot s} \tag{8.39}$$

$$G_{CVC}(s) = \frac{2.178 \cdot s + 47352}{s} \tag{8.40}$$

Figure 8.13 shows the simulation of the complete SMC-PI controller based on Ψ_{i_C} regulating the PV voltage. This simulation considers the same large perturbations used in the previous examples, where, similarly to the SMC based on Ψ_{i_L}, load perturbations are completely rejected. However, due to the complete control of the PV voltage derivative, perturbations in the irradiance are much more mitigated. In addition, the reference voltage provided by an MPPT algorithm is also accurately tracked.

Finally, this PV voltage regulator based on Ψ_{i_C} gives the system stability and provides improved performance over other regulators, both in terms of perturbation rejection

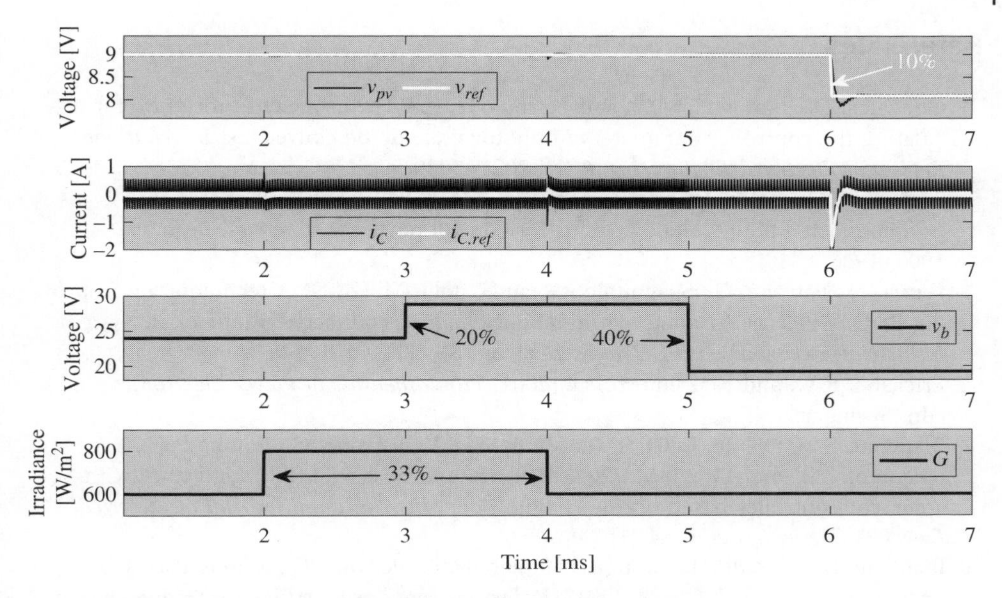

Figure 8.13 Simulation of PV voltage control including the SMC for Ψ_{i_C}.

and reference tracking. Moreover, since this voltage regulator does not require parameterization around a particular operating point, the control performance is the same for all of the operating range. The main drawbacks of this final controller, in comparison with the SMC based on Ψ_{i_L}, are:

- higher complexity in the mathematical analysis
- it is not stable since it requires a voltage regulator to define the operating conditions
- both positive and negative current values must be measured.

8.4 Summary

This chapter has presented design procedures for linear and non-linear control techniques aimed at improving the performance of PV systems.

The two methods introduced for controlling the PV voltage of the MIU were a linear approach and a sliding-mode technique. The first method is based on the linear models introduced in Section 7.4, where poles and zeros are added to fulfill the desired performance criteria. The second method aims to ensure the stability of the MIU voltage in any operational condition despite the non-linear behavior of the circuit. The non-linear control method is robust but its analysis is more complex. Moreover, the implementation hardware exhibits variable switching frequency, which introduces problems related to noise filtering. Switched models can be used to determine the operational conditions in order to define safe implementation hardware. Finally, the control techniques have been tested with predefined reference values to illustrate their performance, and those techniques are able to operate in agreement with an MPPT algorithm as reported in the literature [3, 5, 6].

References

1 Gonzalez, D., Ramos-Paja, C., and Petrone, G. (2011) Automated procedure for calculating the controller parameters in photovoltaic dc/dc converters. *International Review of Electrical Engineering*, **6** (7), 3027–3040.

2 Trejos, A., Gonzalez, D., and Ramos-Paja, C.A. (2012) Modeling of step-up grid-connected photovoltaic systems for control purposes. *Energies*, **5** (6), 1900–1926.

3 Femia, N., Petrone, G., Spagnuolo, G., and Vitelli, M. (2009) A technique for improving P&O MPPT performances of double-stage grid-connected photovoltaic systems. *Industrial Electronics, IEEE Transactions on*, **56** (11), 4473–4482.

4 Erickson, R.W. and Maksimovic, D. (2001) *Fundamentals of Power Electronics*, 2nd edn. Springer.

5 Bianconi, E., Calvente, J., Giral, R., Mamarelis, E., Petrone, G., Ramos-Paja, C.A., Spagnuolo, G., and Vitelli, M. (2013) Perturb and observe MPPT algorithm with a current controller based on the sliding mode. *International Journal of Electrical Power & Energy Systems*, **44** (1), 346–356.

6 Bianconi, E., Calvente, J., Giral, R., Mamarelis, E., Petrone, G., Ramos-Paja, C., Spagnuolo, G., and Vitelli, M. (2013) A fast current-based MPPT technique employing sliding mode control. *Industrial Electronics, IEEE Transactions on*, **60** (3), 1168–1178.

7 Ogata, K. (2005) *Modern Control Engineering*, 3rd edn. Prentice Hall.

8 MathWorks (2014) *sisotool: Interactively design and tune SISO feedback loops*. URL: http://www.mathworks.com/help/control/ref/sisotool.html.

9 Femia, N., Petrone, G., Spagnuolo, G., and Vitelli, M. (2005) Optimization of perturb and observe maximum power point tracking method. *Power Electronics, IEEE Transactions on*, **20** (4), 963–973.

10 Tan, S.C., Lai, Y.M., and Tse, C.K. (2011) *Sliding Mode Control of Switching Power Converters: Techniques and Implementation*, 1st edn. CRC Press.

11 Sira-Ramirez, H. (1987) Sliding motions in bilinear switched networks. *Circuits and Systems, IEEE Transactions on*, **34** (8), 919–933.

12 Rossetto, L., Spiazzi, G., Tenti, P., Fabiano, B., and Licitra, C. (1994) Fast-response high-quality rectifier with sliding mode control. *IEEE Transactions on Power Electronics*, **9** (2), 146–152.

13 Eicker, U. (2003) *Solar Technologies for Buildings*, Wiley, 1st edn.

Index

Photovoltaic Sources Modeling, First Edition. Giovanni Petrone, Carlos Andrés Ramos-Paja and Giovanni Spagnuolo.

Companion Website: www.wiley.com/go/petrone/Photovoltaic_Sources_Modeling